N. Nedjah, E. Alba, L. de Macedo Mourelle (Eds.)

Parallel Evolutionary Computations

Studies in Computational Intelligence, Volume 22

Editor-in-chief
Prof. Janusz Kacprzyk
Systems Research Institute
Polish Academy of Sciences
ul. Newelska 6
01-447 Warsaw
Poland
E-mail: kacprzyk@ibspan.waw.pl

N. Nedjah
E. Alba
L. de Macedo Mourelle
(Eds.)

Parallel Evolutionary Computations

 Springer

Dr. Nadia Nedjah
Department of System Engineering
and Computation
Faculty of Engineering
State University of Rio de Janeiro
Rua São Francisco Xavier, 524, 50. andar
Maracanã, CEP 20559-900
Rio de Janeiro, RJ
Brazil
E-mail: nadia@eng.uerj.br

Professor Enrique Alba
Universidad de Málaga
Depto. of Lenguajes y Ciencias de la
Computación
Campus de Teatinos
29071 Málaga, Spain
E-mail: eat@lcc.uma.es

Dr. Luiza de Macedo Mourelle
Department of System Engineering
and Computation
Faculty of Engineering
State University of Rio de Janeiro
Rua São Francisco Xavier, 524, 50. andar
Maracanã, CEP 20559-900
Rio de Janeiro, RJ
Brazil
E-mail: ldmm@eng.uerj.br

ISSN print edition: 1860-949X
ISSN electronic edition: 1860-9503

ISBN 978-3-642-06939-0
e-ISBN 978-3-540-32839-1

Springer is a part of Springer Science+Business Media
springer.com
© Springer-Verlag Berlin Heidelberg 2006
Softcover reprint of the hardcover 1st edition 2006

To the memory of my father Ali and my beloved mother Fatiha,

Nadia Nedjah

Thanks to my wife Ana,

Enrique Alba

To the memory of my beloved father Luiz and my mother Neuza,

Luiza Mourelle

Preface

Evolutionary algorithms and their related computations are solver systems which use computational models inspired in Darwinian natural selection processes as a key element in their design and implementation. In general, evolutionary computation (EC) is used to solve NP-hard problems which cannot be solved with other tools because of their intrinsic difficulty, high dimensionality or incomplete definition. In practice, EC is composed of a set of different families of algorithms that iteratively improve a set of tentative solutions to obtain an optimal or quasi-optimal solution to the problem. Therefore, evolutionary algorithms require sometimes a massive computational effort to yield efficient and competitive solutions to real-size engineering problems which would otherwise rest unsolved today.

This book focuses on the aspects related to the parallelization of evolutionary computations, such as parallel genetic operators, parallel fitness evaluation, distributed genetic algorithms, and parallel hardware implementations, as well as on the impact of parallel EC on several applications.

The book is divided into four parts. The first part deals with a clear software-like and algorithmic vision on parallel evolutionary optimizations. The second part is about hardware implementations of genetic algorithms, a valuable topic which is hard to find in the present literature. The third part treats the problem of distributed evolutionary computation and presents three interesting applications wherein parallel EC new ideas are featured. Finally, the last part deals with the up-to-date field of parallel particle swarm optimization to illustrate the intrinsic similarities and potential extensions to techniques in this domain.

The goal of this volume has been to offer a wide spectrum of sample works developed in leading research throughout the world about parallel implementations of efficient techniques at the heart of computational intelligence. The book should be useful both for beginners and experienced researchers in the field of computational intelligence.

Part I: Parallel Evolutionary Optimization

In Chapter 1, which is entitled *A Model for Parallel Operators in Genetic Algorithms*, the authors analyze a model for applying crossover and varying mutation separately in parallel. The authors focus on the gains of performance that can be achieved from the concurrent application of variation operators with different and complementary roles. An important aim of this chapter is to show that, for the actual parallelization of operators to be meaningful, appropriate models for the concurrent application of operators should be devised first.

In Chapter 2, which is entitled *Parallel Evolutionary Multiobjective Optimization*, the authors' first goal is to provide the reader with a wide overview of the literature on parallel evolutionary algorithms for multiobjective optimization. Also, the authors include an experimental study wherein *p*PAES, a parallel evolutionary algorithm for multiobjective optimization based on the Pareto Archived Evolution Strategy (PAES) is developed and analyzed. The obtained results show that *p*PAES is a promising option for solving multiobjective optimization problems.

Part II: Parallel Hardware for Genetic Algorithms

In Chapter 3, which is entitled *A Reconfigurable Parallel Hardware for Genetic Algorithms*, the authors propose a massively parallel architecture for hardware implementation of genetic algorithms. The authors claim that this design is quite innovative as it provides a viable solution to the fitness computation problem, which depends heavily on the problem-specific knowledge. The proposed architecture is completely independent of such specifics. It implements the fitness computation using a neural network. The hardware implementation of the neural network is stochastic and thus minimises the required hardware area without much increase in response time. The authors also demonstrate the characteristics of the proposed hardware and compare it to existing ones.

In Chapter 4, which is entitled *Reconfigurable Computing and Parallelism for Implementing and Accelerating Evolutionary Algorithms*, the authors present two instances of hardware implementation of evolutionary algorithms. Both instances provide distinct points of view on how to apply reconfigurable computing technology to increase algorithm efficiency. The cases considered are genetic algorithms and one parallel evolutionary algorithm. In particular, the authors illustrate through an experimental study the performance of the presented hardware implementations using the Travelling Salesman Problem.

Part III: Distributed Evolutionary Computation

In Chapter 5, which is entitled *Performance of Distributed GAs on DNA Fragment Assembly*, the authors present results on analyzing the behaviour of a parallel distributed genetic algorithm over different LAN technologies. Their main goal is to offer a study on the potential impact in the search mechanics when shifting between LANs: a Fast Ethernet network, a Gigabit Ethernet network, and a Myrinet network. They also study the importance of several parameters of the migration policy. The author use the DNA fragment assembly problem to show the actual power and utility of the proposed distributed technique.

In Chapter 6, which is entitled *On Parallel Evolutionary Algorithms on the Computational Grid*, the authors analyze the major traditional parallel models of evolutionary algorithms. The objective consists of adapting the parallel models to to grids taking into account the characteristics of such execution environments in terms of volatility, heterogeneity, large scale and multi-administrative domain. The authors give an overview of such frameworks and present a case study related to ParadisEO-CMW which is a porting of ParadisEO onto Condor and MW allowing a transparent deployment of the parallel EAs on grids.

In Chapter 7, which is entitled *Parallel Evolutionary Algorithms on Consumer-Level Graphics Processing Unit*, the author propose to implement a parallel evolutionary algorithm on consumer-level Graphics Processing Unit. They perform experiments to compare the proposed parallel evolutionary algorithm with an ordinary one and demonstrate that the former is much more effective than the latter. The authors claim that as consumer-level graphics processing units are already widely available and installed on personal computers and they are easy to use and manage, more people will be able to use our parallel algorithm to solve their problems encountered in real-world applications.

Part IV: Parallel Particle Swarm Optimization

In Chapter 8, which is entitled *Intelligent Parallel Particle Swarm Optimization Algorithms*, the authors present a parallel version of the particle swarm optimization (PPSO) algorithm together with three communication strategies which can be used according to the independence of the data. They discuss some communication strategies for PPSO, which can be used according to the strength of the correlation of parameters. Experimental results confirm the superiority of the PPSO algorithms.

In Chapter 9, which is entitled *Parallel Ant Colony Optimization for 3D Protein Structure Prediction using the HP Lattice Model*, the authors introduce a novel method of solving the Hydrophobic-Hydrophilic (HP) protein folding problem in both two and three dimensions using Ant Colony Optimizations and a distributed programming paradigm. They claim that tests across a small number of processors indicate that the multiple colony distributed ACO (MACO) approach outperforms single colony implementations and that experimental results also demonstrate that the proposed algorithms perform well in terms of network scalability.

We are very much grateful to the authors of this volume and to the reviewers for their tremendous service by critically reviewing the chapters. The editors would also like to thank Prof. Janusz Kacprzyk, the editor-in-chief of the Studies in Computational Intelligence Book Series and Dr. Thomas Ditzinger from Springer-Verlag, Germany for their editorial assistance and excellent collaboration to produce this scientific work. We hope that the reader will share our excitement on this **Parallel Evolutionary Computations** volume and will find it useful.

Brazil, Spain
January 2006

Nadia Nedjah
Enrique Alba
Luiza M. Mourelle

Contents

Part II Parallel Hardware for Genetic Algorithms

Part III Distributed Evolutionary Computation

5 Performance of Distributed GAs on DNA Fragment Assembly

6 On Parallel Evolutionary Algorithms on the Computational Grid

Part IV Parallel Particle Swarm Optimization

List of Figures

List of Tables

List of Algorithms

Part I

Parallel Evolutionary Optimization

A Model for Parallel Operators
in Genetic Algorithms

Hernán Aguirre and Kiyoshi Tanaka

Faculty of Engineering, Shinshu University
4-17-1 Wakasato, Nagano, 380-8553 JAPAN
hernan@cs.shinshu-u.ac.jp

In this chapter we analyze a model for applying parallel operators in Genetic Algorithms. Here we focus on crossover and varying mutation applied separately in parallel, emphasizing the gains on performance that can be achieved from the concurrent application of operators with different and complementary roles. We analyze the model in single population Genetic Algorithms using deterministic, adaptive, and self-adaptive mutation rate controls and test its performance on a broad range of classes of 0/1 multiple knapsack problems. We compare the proposed model with the conventional one, where varying mutation is also applied after crossover, presenting evidence that varying mutation parallel to crossover gives an efficient framework to achieve higher performance. We also show that the model is superior for online adaptation of parameters and contend that it is a better option for co-adaptation of parameters. In addition, we study the performance of parallel operators within distributed Genetic Algorithms showing that the inclusion of varying mutation parallel to crossover can increase considerably convergence reliability and robustness of the algorithm, reducing substantially communication costs due to migration.

1.1 Introduction

The development of parallel implementations of algorithms has been mainly motivated by the desire to reduce the overall time to completion of a task by distributing the work implied by a given algorithm to processing elements working in parallel [1]. An alternative approach explores parallel computational models that can exploit interactions among primitive components in order to induce synergetic behaviors for the entire system [2].

There are a variety of models for parallelizing Genetic Algorithms (GAs) in the evolutionary algorithms literature. They have been separated in four

H. Aguirre and K. Tanaka: *A Model for Parallel Operators in Genetic Algorithms*, Studies in Computational Intelligence (SCI) **22**, 3–31 (2006)
www.springerlink.com   © Springer-Verlag Berlin Heidelberg 2006

main categories: global master-slave, island, cellular, and hierarchical parallel GAs [3, 4, 5, 6]. In a global master-slave GA there is a single population and the evaluation of fitness is distributed among several processors. Selection, crossover and mutation consider the entire population [7]. An island GA, also known as coarse-grained or distributed GA, consists on several subpopulations evolving separately with occasional migration of individuals between subpopulations [1, 8, 9]. A cellular or fine-grained GA consists of one spatially structured population. Selection and mating are restricted to a small neighborhood. The neighborhoods are allowed to overlap permitting some interaction among individuals [10, 11]. Finally, a hierarchical parallel GA combines an island model with either a master-slave or cellular GA [5].

The global master-slave GA does not affect the behavior of the algorithm and can be considered only as a hardware accelerator. However, the other parallel formulations of GAs are very different from canonical GAs [12, 13], especially with regards to population structure and selection mechanisms. These modifications change the way the GA works affecting its dynamics and the trajectory of evolution. For example, subpopulation size, migration rate, and migration frequency are crucial to the performance of island models. Cellular, island and hierarchical models perform as well as or better than canonical versions and have the potential of being more than just hardware accelerators [3, 4, 5].

Another aspect of GAs that can be parallelized is the application of crossover and mutation. However, these operators from a processing time stand are usually simple and their parallelization has been mostly overlooked precisely because any gain we might expect reducing the overall time to completion of the algorithm would seem minor. The processing time viewpoint alone misses the dynamics that can arise from operators with complementary roles acting in parallel. We argue that rather than as a hardware accelerator, the more significant gains from the parallel application of operators could come from exploiting in a better way the interaction between them.

The balance between crossover and mutation is crucial to the performance of GAs. One way to pursue better balances, and therefore better performance, is to combine crossover with varying mutation rates. The conventional model tries to achieve a better mix for crossover and varying mutation applying one operator after the other, as in canonical GAs, and including a selection method with strong selection pressure. Approaches based on this model have been shown to be successful compared to simple GAs that apply constant small mutation rates and could be candidates for actual parallel implementation. Yet, these approaches in principle are prone to interferences between operators and condemned to be inefficient, misleading for parameters adaptation, and not likely to be useful for parameters co-adaptation.

In this chapter we analyze a model for applying crossover and varying mutation separately in parallel. Here we focus on the gains on performance that can be achieved from the concurrent application of variation operators with different and complementary roles. Similar to distributed and cellular

GAs, the parallelization of operators could also affect the behavior of the algorithm and change its dynamics. An important aim of this work is to show that, for the actual parallelization of operators to be meaningful, appropriate models for the concurrent application of operators should be devised first. Although the benefits highlighted here can be achieved by concurrent process without actually implementing them in parallel, the ultimate goal of this work is to have a model for parallel operators that would work appropriately once implemented in parallel hardware architectures where the effects of concurrence must be dealt with. This work is an important step towards a better understanding of the interactions between concurrent crossover and varying mutations, necessary to design and build parallel evolutionary algorithms with parallel operators. Present parallel implementations of distributed and cellular evolutionary algorithms benefit from a well established body of research on the effects of different population structures and selection. We need also to develop the basis for understanding better the concurrent application of operators. This work is a contribution on those lines.

In our analysis we include, besides an adaptive mutation rate control proposed by the authors, two well known deterministic and self-adaptive mutation rate controls methods in order to study the conventional and the proposed model for applying crossover and varying mutations in parallel. The motivation for this is that we want to show that differences in performance are not due to an especial or particular mutation rate control, but rather consistently derive from the model for applying crossover and varying mutations separately in parallel. We analyze the model in single population and distributed GAs presenting evidence that the proposed model for parallel operators gives an efficient framework to achieve high performance by keeping the strengths of the individual operators without interfering one with the other. We also show that the model is superior for online adaptation of parameters and contend that it is a better option for co-adaptation of parameters. In the future we would like to implement this model on a parallel architecture, where in addition to dealing with the effects of concurrent process, we will need to cope with the demands proper of parallel implementations.

The rest of the chapter is organized as follows. Section 1.2 describes from a parallel and concurrent operators stand point the way mutation and crossover are used in canonical GAs and conventional varying mutation GAs. Section 1.3 presents the proposed model for varying mutations parallel to crossover. Section 1.4 describes 0/1 multiple knapsack problems and the instances used to test and compare the proposed model. Section 1.5 studies the effectiveness of the model using adaptive mutations and section 1.6 compares its performance against canonical GAs and the conventional varying mutation GAs using deterministic, and self-adaptive mutation rate controls. Section 1.7 discusses the effectiveness of the application of parallel operators within distributed GAs. Finally, section 1.8 presents conclusions.

1.2 Implicit Parallel Operators in Canonical and Conventional Varying Mutation GAs

A canonical GA [12, 13] selects individuals from the parent population $P(t)$ with a selection probability proportional to their fitness and applies crossover with probability p_c followed by mutation with a very small constant mutation probability p_m per bit, i.e. background mutation. In the absence of crossover $(1 - p_c)$ mutation is applied alone. Conventional varying mutation GAs differ from canonical GAs mainly in the mutation rate control. Also, some of them use a selection mechanism with selection pressure higher than the canonical GA. However, the application of operators has been similar to canonical GAs.

From the application of operators standpoint, it can be said that within canonical GAs and conventional varying mutation GAs the probability of crossover p_c enables an implicit parallel application of two operators. One of the operators is crossover followed by mutation (CM) and the other one is mutation alone (M). It should be noted that mutation in both CM and M is governed by the same mutation probability p_m and applies the same *bit by bit* mutation strategy. The number of offspring created by CM and M, $\lambda_{CM}^{(t)}$ and $\lambda_{M}^{(t)}$, respectively, depends on the probability of crossover p_c and may vary at each generation t due to the stochastic process. However, the total number of offspring λ remains constant. Fig. 1.1 presents a block diagram of the canonical GA and illustrates the implicit parallel application of CM and M with constant small mutation. Similarly, Fig. 1.2 illustrates the implicit parallel application of CM and M when varying mutations are used.

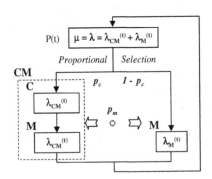

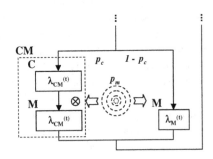

Fig. 1.1. Canonical Genetic Algorithm

Fig. 1.2. Genetic Algorithm with Conventional Varying Mutation

Since the probability p_c is usually set to 0.6, and higher values are often used [14], it turns out that mutation is mostly applied serial to crossover. In canonical GAs p_m is small, therefore the amount of diversity introduced by mutation either through CM or M is modest. For the same reason, the disruption that mutation causes to crossover in CM is also expected to be small. In varying mutation GAs, however, mutations are higher than canonical GAs and the combined effect of crossover and mutation in CM as well as the effect of mutation alone in M should be carefully reconsidered.

In the case of CM, besides those cases in which crossover and mutation aggregate in a positive manner or are neutral, those cases in which one of the operators is working well but is being hampered by the other should also be taken into account. For example, if mutation probabilities were high then although crossover could be doing a good job it is likely that some of the just created favorable recombinations would be lost, before they become fixed in the offspring, due to the high disruption introduced by mutation. We can think of this case as a mutation interference with crossover in the *creation* of beneficial recombinations. On the other hand, mutation could be working well but crossover may produce poor performing individuals affecting the survivability of beneficial mutations that can contribute to the search. We can think of this case as a crossover interference with mutation in the introduction of beneficial mutations.

Important questions in the model of conventional varying mutation GAs are how these interferences between crossover and higher mutation in CM affect the algorithm. Does it affect performance of the algorithm expressed in terms of convergence speed and reliability? Does it affect the mutation rate control itself? Is it an appropriate model for combining forms of control (co-adaptation of parameters)?

1.3 A Model of Parallel Varying Mutation GA (GA-SRM)

To answer the questions raised by the standard varying mutation approach, this work explores a model of GA that explicitly differentiate the mutation operator applied parallel to crossover from the mutation operator applied after crossover. There are some advantages to this differentiation. (1) Each mutation operator could be assigned its own mutation probability. Thus, varying mutation can be applied only parallel to crossover and mutation after crossover can be applied with a small mutation rate (or none at all) avoiding interferences between crossover and high mutation in CM. (2) The parallel varying mutation operator does not diminish the effectiveness of other operator. In this case, high mutations when are harmful will have a negative impact on the *propagation* of beneficial recombinations already present in the parent population. However, it will not affect the *creation* of beneficial recombinations by crossover as high mutation can do it in CM. (3) Since the instantaneous effectiveness of the varying mutation operator depends only upon itself its relative success can be directly tied to the mutation rate to create effective adaptive and self-adaptive schemes for mutation rate control. (4) Parallel mutation can be studied on its own. For example, higher mutation rates raise the question of mutation strategy and its relevance to a given epistatic pattern or class of problems. (5) The individual roles and the interaction of crossover and varying mutation throughout the run of the algorithm can be better understood, which could be important for co-adaptation studies.

The main components of the proposed model are the application of varying mutation exclusively parallel to crossover and an extinctive selection mechanism. We study the model using deterministic, adaptive, and self-adaptive mutation rate controls. We also look into the importance of mutation strategy to choose the bits that will undergo mutation in the parallel varying mutation operator. Fig. 1.3 illustrates a parallel varying mutation GA based on the proposed model. In the following we present the details of the model.

1.3.1 Explicit Parallel Operators

The two operators included in the model are crossover followed by background mutation (CM) and varying mutation parallel to CM [15, 16, 17]. To clearly distinguish between the mutation operator applied after crossover and the mutation operator applied parallel to crossover, the parallel varying mutation operator is called *Self-Reproduction with Mutation* (SRM). In the following this model of GA is called GA-SRM.

As suggested above, the explicit parallel formulation of CM and SRM could give an efficient framework to achieve better balances for mutation and crossover during the run of the algorithm in which the strengths of higher mutation and crossover can be kept without interfering one with the other. SRM parallel to CM would implicitly increases the levels of cooperation to introduce beneficial mutations and create beneficial recombinations. It also sets the stage for competition between operators' offspring. The number of offspring λ_{CM} and λ_{SRM} that will be created by CM and SRM, respectively, is deterministically decided at the beginning of the run.

1.3.2 Extinctive Selection

Recent works have given more insights to better characterize the roles of recombination and mutation in evolutionary algorithms [18]. An important issue to consider is the deleterious effects caused by the operators and especially how to deal with them. The parallel formulation of CM and SRM can avoid interferences between crossover and high mutation; however it cannot prevent SRM from creating deleterious mutations or CM from producing ineffective crossing over operations. To cope with these cases the model also incorporates the concept of extinctive selection that has been widely used in Evolution Strategies. Through extinctive selection the offspring created by CM and SRM coexist competing for survival (the poor performing individuals created by both operators are eliminated) and reproduction.

Among the various extinctive selection mechanisms available in the EA literature (μ, λ) proportional selection [19] is chosen to implement the required extinctive selection mechanism. Selection probabilities are computed by

$$P_s(x_i^{(t)}) = \begin{cases} \dfrac{f(x_i^{(t)})}{\displaystyle\sum_{j=1}^{\mu} f(x_j^{(t)})} & (1 \leq i \leq \mu) \\[1em] 0 & (\mu < i \leq \lambda) \end{cases} \tag{1.1}$$

where $x_i^{(t)}$ is an individual at generation t which has the i-th highest fitness value $f(x_i^{(t)})$, μ is the number of parents and λ is the number of offspring. The parallel formulation of genetic operators tied to extinctive selection creates a cooperative-competitive environment for the offspring created by CM and SRM.

1.3.3 Mutation Rate Control

Deterministic, adaptive, and self-adaptive mutation rate control schedules are used to study the parallel application of varying mutations. Deterministic and adaptive schedules use only one mutation rate for all the individuals in the population whereas the self-adaptive schedule uses one mutation rate per individual.

Deterministic

The deterministic approach implements a time-dependent mutation schedule that reduces mutation rate in a hyperbolic shape [20] and is expressed by

$$p_m^{(t)} = \left(r_o + \frac{n - r_o}{T - 1} t \right)^{-1} \tag{1.2}$$

where T is the maximum number of generations, $t \in \{0, 1, \cdots, T - 1\}$ is the current generation, and n is the bit string length. The mutation rate $p_m^{(t)}$ varies in the range $[1/r_o, 1/n]$. In the original formulation $r_o = 2$. Here r_o is included as a parameter in order to study different ranges for mutation. In the deterministic approach the mutation rate calculated at time t is applied to all individuals created by SRM. Fig. 1.4 illustrates the mutation rates over the generations by this schedule for three initial mutation rates, $p_m^{(0)} = \{0.5, 0.10, 0.05\}$.

Adaptive

Since the instantaneous effectiveness of SRM depends only upon itself its relative success can be directly tied to the mutation rate to create effective adaptive and self-adaptive schemes for mutation rate control. In this work the adaptive mutation rate control dynamically adjusts the mutation rate within

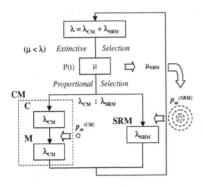

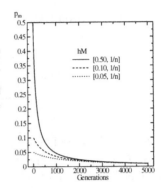

Fig. 1.3. GA with Parallel Varying Mutation

Fig. 1.4. Deterministic Hyperbolic Schedule for Mutation Rate Control

SRM every time a normalized mutants survival ratio γ falls under a threshold τ [15] (see 1.3.4). The ratio γ is specified by

$$\gamma = \frac{\mu_{SRM}}{\lambda_{SRM}} \cdot \frac{\lambda}{\mu} \tag{1.3}$$

where μ_{SRM} is the number of individuals created by SRM present in the parent population $P(t)$ after extinctive selection at time t, λ_{SRM} is the number of offspring created by SRM, λ is the total number of offspring ($\lambda_{CM} + \lambda_{SRM}$), and μ is the number of individuals in $P(t)$.

Self-Adaptive

To include self-adaptation, each individual incorporates its own mutation probability within the representation. SRM to produce offspring first mutates the mutation probability of the selected individual and then mutates the object variable using the individual's mutated probability. This work applies a self-adaptive approach that uses a continuous representation for the mutation rate [20] and mutates the mutation probability of each individual by

$$p_m^{(t)}(i) = \left(1 + \frac{1 - p_m^{(t-1)}(i)}{p_m^{(t-1)}(i)} exp(-\gamma N(0,1)) \right)^{-1} \tag{1.4}$$

where i indicates the i-th individual, γ is a learning rate that control the speed of self-adaptation, and $N(0,1)$ is a normally distributed random number with expectation zero and standard deviation one. Note that individuals selected to reproduce with SRM at generation t could have been created either by SRM or CM at generation $t-1$. Since the mutation rate of each individual is mutated only by SRM, individuals created by CM do not carry an updated mutation rate. Thus, the mutation rate of individuals that were created by CM at generation $t-1$ is first updated by

$$p_m^{(t-1)}(j) = \frac{1}{\mu_{SRM}} \sum_{k=1}^{\mu_{SRM}} p_m^{(t-1)}(k) \qquad (1.5)$$

where j indicates an individual created by CM at $(t-1)$, k indicates the individuals created by SRM at $(t-1)$ that survived extinctive selection, and μ_{SRM} is the number of offspring created by SRM that survived extinctive selection. In the case that no offspring created by SRM survived extinctive selection, $p_m^{(t-1)}(j)$ is set to the mutation value of the best SRM's offspring. SRM will mutate this updated mutation in order to mutate the object variable.

1.3.4 Mutation Strategy

In the case of background mutation we expect on the average to flip 1 bit (or less) in each individual at each generation. When higher mutations are applied, however, many more bits would be flipped in the same individual. This raises the question of whether a mutation strategy to choose the bits that will undergo mutation would be more effective than other and for which classes of problems. To study this point, two mutation strategies are investigated for SRM: (i) adaptive dynamic-segment (ADS), and (ii) adaptive dynamic-probability (ADP). Both schemes, ADS and ADP, impose an adaptive mutation rate control with the same expected average number of flipped bits; the difference lies whether mutation is applied locally inside a mutation segment (ADS) or globally inside the whole bit string (ADP).

Adaptive Dynamic-Segment (ADS)

ADS mutates within a segment of the chromosome with constant probability $p_m^{(SRM)} = \alpha$ per bit varying the length ℓ of the mutation segment each time the mutants survival ratio γ falls under a threshold τ. At time $t = 0$, $\ell = n$ (bit string length). Afterwards, if $(\gamma < \tau)$ and $(\ell > 1/\alpha)$ then $\ell = \ell/2$. The segment initial position, for each individual, is chosen at random $s_i = N[0, n)$ and its final position is calculated by $s_f = (s_i + \ell) \bmod n$. With this scheme, the segment size ℓ varies from n to $1/\alpha$ and the average number of flipped bits goes down from $n\alpha$ to 1.

Adaptive Dynamic-Probability (ADP)

ADP mutates every bit in the chromosome with probability $p_m^{(SRM)}$ varying each time γ falls under τ. At time $t = 0$, $p_m^{(SRM)} = \alpha$. Afterwards, if $(\gamma < \tau)$ and $(p_m^{(SRM)} > 1/n)$ then $p_m^{(SRM)} = p_m^{(SRM)}/2$. In other words, the mutation segment lenght is kept constant $\ell = n$, but $p_m^{(SRM)}$ follows a step decreasing approach from α to $1/n$ per bit.

1.4 0/1 Multiple Knapsacks Problems

The 0/1 multiple knapsack problem is a NP-hard combinatorial optimization problem and its importance is well known both from a theoretic and practical point of view. In the 0/1 multiple knapsack problem there are m knapsacks and n objects. The capacities of the knapsacks are $c_1, c_2, ..., c_m$. For each object there is a profit p_i $(1 \leq i \leq n)$ and a set of weights w_{ij} $(1 \leq j \leq m)$, one weight per knapsack. If an object is selected its profit is accrued and the knapsacks are filled with the objects' weights. The problem consists on finding the subset of objects that maximizes profit without overfilling any of the knapsacks with objects' weights. The 0/1 multiple knapsack problem can be formulated to maximize the function

$$g(x) = \sum_{i=1}^{n} p_i x_i \qquad (1.6)$$

subject to

$$\sum_{i=1}^{n} w_{ij} x_i \leq c_j \qquad (j = 1, ..., m) \qquad (1.7)$$

where $x_i \in \{0,1\}$ $(i = 1, ..., n)$ are elements of a solution vector $x = (x_1, x_2, ..., x_n)$, which is the combination of objects we are interested in finding. Solutions to this problem have a natural binary representation in GAs constructed by mapping each object to a locus within the binary chromosome. A 1 in locus i indicates that the object i is being selected and a 0 otherwise. A solution vector x should guarantee that no knapsack is overfilled and the best solution should yield the maximum profit. An x that overfills at least one of the knapsacks is considered as an infeasible solution. Fig. 1.5 illustrates the problem.

Besides defining the number of knapsacks m (number of constraints) and the number of objects n (size of the search space, 2^n), it is also possible to define the tightness ratio ϕ between knapsack capacities and object weights, which implies a ratio between the feasible region and the whole search space. To obtain a broad perspective on performance and scalability the GAs are applied to several subclasses of knapsacks problems systematically varying ϕ, m, and n. We use 7 subclasses as shown in Table 1.1 and 10 random problems for each subclass. These allow to observe the robustness of the algorithms reducing the feasible region $(\phi = \{0.75, 0.50, 0.25\})$, increasing the number of constraints $(m = \{5, 10, 30\}$ knapsacks), and increasing the search space $(n = \{100, 250, 500\}$ objects). The set of 0/1 multiple knapsacks problems selected to test the algorithms was obtained from the OR-Library[1].

The quality of the solutions found by the algorithms are measured by the average percentage error gap in a subclass of problems, which is calculated as the normalized difference between the best solutions found and the optimal

---

[1] http://mscmga.ms.ic.ac.uk/jeb/orlib/info.html

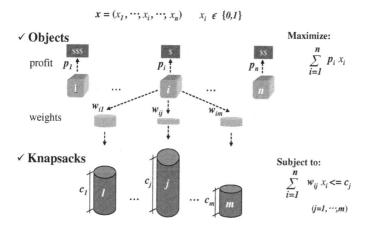

$$x = (x_1, \cdots, x_i, \cdots, x_n) \qquad x_i \in \{0,1\}$$

✓ **Objects**

profit p_1

p_i

p_n

weights

w_{i1} w_{ij} w_{im}

✓ **Knapsacks**

c_1 c_j c_m

Maximize:

$$\sum_{i=1}^{n} p_i x_i$$

Subject to:

$$\sum_{i=1}^{n} w_{ij} x_i <= c_j$$

$$(j=1, \cdots, m)$$

Fig. 1.5. Knapsack Problem

value given by the linear programming relaxation (LP) (the optimal integer solutions are unknown) [23]. The average error gap is taken for the 10 random problems in each subclass performing 50 runs for each problem by

$$\% \ Error \ Gap = \frac{1}{10} \sum_{p=1}^{10} 100 \times \frac{LP - (Problem \ Average)_p}{LP}$$

$$Problem \ Average = \frac{1}{50} \sum_{r=1}^{50} (Best \ Solution)_r$$

To deal with infeasible solutions a penalty term is introduced into the fitness function f as follows

$$f(\boldsymbol{x}) = \begin{cases} g(\boldsymbol{x})/s \cdot max\{o_j\} & (s > 0) \\ g(\boldsymbol{x}) & (s = 0) \end{cases} \qquad (1.8)$$

where s $(0 \leq s \leq m)$ is the number of overfilled knapsacks and o_j (> 1) is the overfilling ratio of knapsack j calculated by

$$o_j = \sum_{i=1}^{n} w_{ij} x_i / c_j. \qquad (1.9)$$

Note that the penalty term is a function of both number of violated constraints (s) and distance from feasibility (o_j). Results reported here were obtained using an initial population randomly initialized with a 0.25 probability for 1s.

Table 1.1. 7 subclasses of problems, 10 random problems in each subclass

Subclass	Paremeters			Comment
	ϕ	m	n	
1	0.75	30	100	reducing
2	0.50			feasible
3	0.25			region
4	0.25	5	100	increasing
5		10		number
(3)		30		constraints
(3)	0.25	30	100	increasing
6			250	search
7			500	space

1.5 Studying the Structure
of the Parallel Varying Mutation GA-SRM

This section studies the model of the parallel varying mutation GA-SRM using the adaptive mechanism explained in 1.3.3. It analyzes the contribution of extinctive selection, adaptation, mutation strategy, and the interaction between parallel varying mutation and crossover. Experiments are conducted using the following algorithms : (i) a canonical GA denoted as cGA, (ii) a simple GA with (μ, λ) proportional selection denoted as GA(μ, λ), and (iii) the proposed algorithm denoted as GA-SRM(μ, λ) (CM, SRM, and extinctive proportional selection). In GA-SRM, SRM is used with ADS and ADP mutation strategies setting the threshold τ to 0.64 and 0.54, respectively. These values were sampled in order to choose the ones that produce the overall higher performance on all subclasses of problems. The genetic algorithms run for $T = 5 \times 10^5$ evaluations set with the parameters specified in Table 1.2. It should be mentioned that the crossover rate for the simple GAs is set to 0.6 because it is a frequent setting for this kind of GA. Note also that this crossover rate is 10% larger than the one used in GA-SRM because the offspring creation ratio for CM and SRM is 1:1. The reader is referred to [15, 17] for a detailed study on the components and parameters of the model and its comparison with simple GAs.

1.5.1 Simple GA and Extinctive Selection

Since most varying mutation algorithms combine higher mutations with a kind of extinctive (truncated) selection, it is important to asses the contributions to performance of a higher selection pressure before analyzing varying mutations. Thus, we first observe the effect of extinctive selection on the performance of a simple GA. Fig. 1.6 plots for one of the problems the fitness of the best-so-far individual over the generations by the canonical cGA(100) and a simple GA using $(\mu, \lambda) = \{(15, 100), (50, 100)\}$ populations.

Table 1.2. Genetic algorithms parameters.

Parameter	cGA	$GA(\mu, \lambda)$	$GA - SRM(\mu, \lambda)$
Representation	Binary	Binary	Binary
Selection	Proport.	(μ, λ) Proport.	(μ, λ) Proport.
Scaling	Linear	Linear	Linear
Mating	$(\boldsymbol{x}_i, \boldsymbol{x}_j), \ i \neq j$	$(\boldsymbol{x}_i, \boldsymbol{x}_j), \ i \neq j$	$(\boldsymbol{x}_i, \boldsymbol{x}_j), \ i \neq j$
Crossover	one point	one point	one point
p_c	0.6	0.6	1.0
$p_m^{(CM)}$	$1/n$	$1/n$	$1/n$
$p_m^{(SRM)}$	–	–	$\begin{cases} \alpha = 0.5, \ \ell = [n, 1/\alpha] \ (ADS) \\ \alpha = [0.5, 1/n], \ \ell = n \ (ADP) \\ [p_m^{(t=0)}, 1/n] \ (hM) \\ [p_m^{(t=0)}, 1/n], \gamma = 0.20 \ (sM) \end{cases}$
$\mu : \lambda$	–	$1 : 2$	$1 : 2$
$\lambda_{CM} : \lambda_{SRM}$	–	–	$1 : 1$

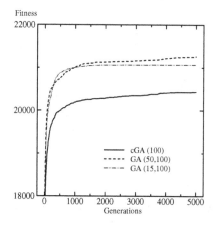

Fig. 1.6. Effect of Extinctive Selection on a Simple GA ($m = 30, n = 100, \phi = 0.25$)

From this figure we see that extinctive selection alone remarkable improves the solution quality reached by the cGA in this kind of problems. As mentioned before, the problems used in this study are highly constrained with sparse feasible regions where algorithms with penalty functions have a hard time finding feasible solutions [22, 23]. A higher selection pressure in these problems is helping the algorithm to focus the search around the feasible regions. Extinctive selection has also another effect. It increases the convergence speed of the algorithm [19]. Both GA(15,100) and GA(50,100) are faster than cGA(100). However, a population of (50,100) gives better final results than

(15,100). In the following we use the results of GA(50,100) as reference for comparison with the varying mutation algorithms.

1.5.2 Contribution of Parallel Genetic Operators and Extinctive Selection

Next, we verify the contribution of the parallel mutation operator SRM and the robustness of its adaptive mechanism. Fig. 1.7 plots the average objective fitness of the best-so-far individual over the generations by GA-SRM(50,100) and GA(50,100). From Fig. 1.7 it can be seen that convergence reliability of GA-SRM(50,100) is much better than GA(50,100). The only difference between GA(μ,λ) and GA-SRM(μ,λ) is the inclusion of parallel adaptive mutation SRM in the latter. Therefore, the increase in performance is directly attributed to SRM and its interaction with CM.

To better observe SRM's contribution experiments are conducted in which starting with a GA(μ,λ) configuration, i.e. all CM and extinctive selection, after a predetermined number of evaluations the algorithm switches to a GA-SRM(μ,λ) configuration, i.e. CM, SRM, and extinctive selection. Fig. 1.7 also includes results by an algorithm that makes the configuration transition from GA(50,100) to GA-SRM(50,100) at $\{0.10T, 0.20T, 0.5T\}$ evaluations, respectively. Every time a transition takes place, initial mutation probability for SRM is set to $p_m^{(SRM)} = 0.5$. From Fig. 1.7 it can be seen that as soon as SRM is included fitness starts to pick up increasing the convergence reliability of the algorithm. Also, note that early transitions produce higher performance. The almost instantaneous pick up on fitness after the transitions in Fig. 1.7 is also an indication of the robustness of the adaptation mechanism. That is, it quickly finds the suitable mutation probability regardless of the stage of the search.

To further clarify the contribution of the interaction between parallel operators CM and SRM during the latest stages of the search experiments are also conducted in which starting with a GA-SRM(μ,λ) configuration, after the mutation rate on SRM has reached a predetermined value, the algorithm switches either to a all CM regime with extinctive selection or to a all SRM regime with extinctive selection. In the latter case no further reductions on SRM's mutation rate are done. Fig. 1.8 plots results by an algorithm that makes the configuration transition as soon as the mutation segment length in SRM has reached $\ell = \{2, 3\}$. Note that since adaptive mutation segment (ADS) strategy is being used and mutation probability is $p_m^{(SRM)} = 0.5$ for the bits within the mutation segment, transitions take place when the mutation rate are in average $1/n$ and $1.5/n$ bits per individual, respectively. As a reference it also includes the results by GA-SRM(50,100). From Fig. 1.8 it can be seen that neither CM nor SRM alone but the parallel interaction of both CM and SRM leads to a higher convergence reliability.

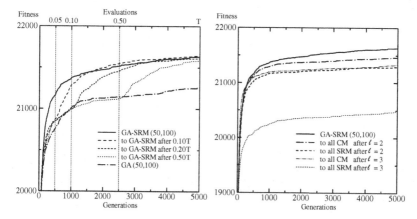

Fig. 1.7. Transition from GA(50,100) to GA-SRM(50,100) at various fractions of T

Fig. 1.8. Transitions from GA-SRM(50,100) to either all CM(50,100) or all SRM(50,100) regimes at $\ell = \{2,3\}$

1.5.3 Results on Classes of 0/1 Multiple Knapsack Problems

To obtain a broader perspective on performance and scalability, we show results on classes of knapsacks problems varying the expected difficulty of the problem by reducing the ratio of the feasible region (ϕ), increasing the number of constraints (knapsacks m), and increasing the number of objects (size of the search space is 2^n). Results are shown in Fig. 1.9, Fig. 1.10, and Fig. 1.11, respectively. Error bars indicate 95% confidence intervals for the mean.

From these figures we can see that GA-SRM algorithms that combine extinctive selection with varying mutations parallel to crossover give significantly better results than a canonical cGA(100) and a simple GA(50,100) with extinctive selection, i.e. smaller error gaps. GA-SRM with ADS is superior to GA-SRM with ADP mutation strategy suggesting that the design of parallel mutation could be an important factor to improve further the search performance of GAs. Note that the difference in performance between GA-SRM with ADS and the other algorithms becomes more apparent for problems considered more difficult (smaller ϕ or larger infeasible regions, larger number of constraints m, and larger search space 2^n). For a detailed study on the effectiveness of mutation strategy and its relation to epistatic patterns among bits the reader is referred to [24].

Another observation is that the performance of GA(50,100) is by far superior to cGA(100) for all values of ϕ, indicating that extinctive selection is an important factor to increase the performance of GAs in problems with infeasible regions. Also, it is worth noting that the quality of solutions found by the GAs decreases, i.e. larger error gaps, as the ratio ϕ is reduced or the number of constraints m increase. This effect is less clear when the search

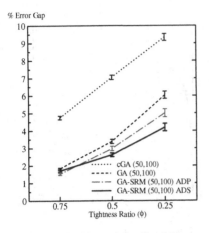

Fig. 1.9. Reducing the feasible region $\phi = \{0.75, 0.50, 0.25\}$, $m = 30$, $n = 100$

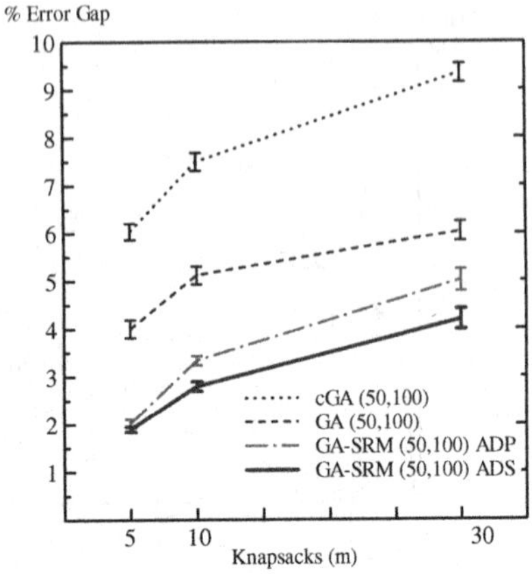

Fig. 1.10. Increasing the number of constraints $m = \{5, 10, 30\}$, $n = 100$, $\phi = 0.25$

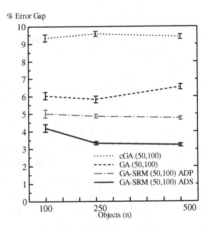

Fig. 1.11. Increasing the search space 2^n, $n = \{100, 250, 500\}$, $m = 30$, $\phi = 0.25$

space becomes larger by increasing the number of objects n, especially in the case of GA-SRM.

1.6 Comparing Conventional and Parallel Varying Mutation Models

This section compares the performance between the conventional and proposed model using deterministic and self-adaptive mutation rate controls for varying mutation [25, 26, 27].

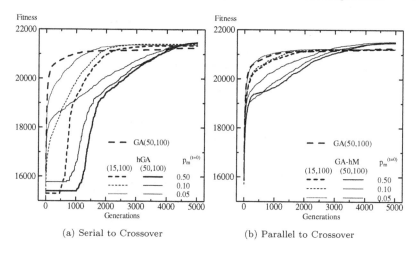

Fig. 1.12. Deterministic Varying Mutation ($m = 30$, $n = 100$, $\phi = 0.25$)

1.6.1 Deterministic Varying Mutation

The hyperbolic schedule of deterministic mutation (1.2) depends only on time and it is not attached to the relative success of mutation. Therefore, it is an ideal candidate to isolate and observe the impact of higher mutations in both models of GAs.

Fig. 1.12 **(a)** and **(b)** plot the average fitness of the best-so-far individual over the generations by a conventional deterministically varying mutation GA (hGA) and a parallel deterministically varying mutation GA (GA-hM), respectively. Results by GA(50,100) are also included for comparison. Experiments are conducted using various populations $(\mu, \lambda) = \{(15, 100), (50, 100)\}$ and initial mutation probabilities $p_m^{(t=0)} = \{0.50, 0.10, 0.05\}$. Except for the mutation rate control, hGA is set with the simple GA(μ,λ) parameters and GA-hM with GA-SRM's parameters as shown in Table 1.2.

From Fig. 1.12 **(a)** it is worth noting the initial *flat* periods in which the fitness of the best-so-far individual in hGA did not improve. This is a clear indication of the disruption caused by high mutation after crossover. Looking at both Fig. 1.12 **(a)** and Fig. 1.12 **(b)** becomes apparent that varying mutation parallel to crossover is less disruptive than varying mutation serial to crossover. Contrary to hGA, in the case of GA-hM there are no initial *flat* periods and in all cases GA-hM converges to higher fitness and faster than hGA for similar values of (μ,λ) and $p_m^{(t=0)}$. Also, as a consequence of this less disruptiveness, the initial value set for varying mutation in GA-hM has a smaller impact on convergence speed than it does in hGA. See for example GA-hM(50,100) for $p_m^{(t=0)} = 0.5$ and $p_m^{(t=0)} = 0.05$ and compare it with hGA for similar settings. Thus, GA-hM is more robust than hGA to initial settings

Table 1.3. Convergence Reliability of Conventional and Parallel Varying Mutation GAs with Deterministic. Mean % error gaps and standard deviations: a) Reducing the feasible region $\phi = \{0.75, 0.50, 0.25\}$, $m = 30$, $n = 100$, b) Increasing the number of constraints $m = \{5, 10, 30\}$, $n = 100$, $\phi = 0.25$, and c) Increasing the search space $n = \{100, 250, 500\}$, $m = 30$, $\phi = 0.25$.

Problem Subclass			Deterministic			
			hGA		GA-hM	
ϕ	m	n	% Error Gap	Stdev	% Error Gap	Stdev
0.75	30	100	1.46	0.10	1.41	0.10
0.50			2.51	0.15	2.42	0.12
0.25			5.07	0.35	4.88	0.40
0.25	5	100	2.39	0.13	2.25	0.16
	10		3.57	0.17	3.41	0.18
	30		5.07	0.35	4.88	0.40
0.25	30	100	5.07	0.35	4.88	0.40
		250	4.18	0.14	4.01	0.16
		500	4.08	0.16	3.88	0.14

Table 1.4. Convergence Reliability of Conventional and Parallel Varying Mutation GAs Self-Adaptive Mutation Rates. Mean % error gaps and standard deviations: a) Reducing the feasible region $\phi = \{0.75, 0.50, 0.25\}$, $m = 30$, $n = 100$, b) Increasing the number of constraints $m = \{5, 10, 30\}$, $n = 100$, $\phi = 0.25$, and c) Increasing the search space $n = \{100, 250, 500\}$, $m = 30$, $\phi = 0.25$.

Problem Subclass			Self-Adaptive			
			sGA		GA-sM	
ϕ	m	n	% Error Gap	Stdev	% Error Gap	Stdev
0.75	30	100	1.84	0.08	1.73	0.10
0.50			3.02	0.21	2.97	0.20
0.25			4.90	0.43	4.40	0.39
0.25	5	100	1.92	0.16	1.95	0.10
	10		3.19	0.18	2.92	0.17
	30		4.90	0.43	4.40	0.39
0.25	30	100	4.90	0.43	4.40	0.39
		250	4.16	0.28	3.61	0.17
		500	3.95	0.13	3.57	0.13

of mutation rate. Also note that in GA-hM, similar to hGA, a $(\mu, \lambda)=(50,100)$ population gives better final results than $(\mu, \lambda)=(15,100)$.

Table 1.3 and Table 1.4 show the average percentage error gap and its standard deviation by hGA and GA-hM. The statistical significance of the results achieved by hGA and GA-hM is verified conducting a two-factors analysis of variance (factorial ANOVA), a procedure for comparing multiple experiments population means. The experiments consist on reducing the feasible region (ϕ), increasing the number of constraints (m), and increasing the search space (n). The factors are the GA type (hGA, GA-hM) and problem difficulty $(\phi = \{0.75, 0.50, 0.25\}$ in the first experiment, $m = \{5, 10, 30\}$ in the second, and $n = \{100, 250, 500\}$ in the third).

From this analysis, the p values corresponding to the GA factor (hGA and GA-hM) in the three experiments are $p = 0.0766$, $p = 0.0157$, and $p = 0.0047$, respectively. The p values are the smallest significant level of error α that would allow rejection of the null hypothesis, i.e. that the means of hGA and GA-hM are the same. These p values reveal that in the case of reducing the feasible region (ϕ) there is *some indication* of an effect by the GA type factor since $p = 0.0766$ is not much greater than $\alpha = 0.05$. In the cases of increasing the number of constraints (m) and the size of the search space (n) the p values 0.0157 and 0.0047, respectively, are considerably less than $\alpha = 0.05$ indicating a *strong main effect* by the GA type concluding that the parallel deterministic varying mutation GA-hM attains significantly smaller error than the conventional deterministic varying mutation hGA.

1.6.2 Self-Adaptive Varying Mutation

A self-adaptive scheme uses one mutation rate per individual, which are usually set at $t = 0$ to random values in the range allowed for mutation. Two important ingredients of self-adaptation are the diversity of parameter settings and the capability of the method to adapt the parameters. It has been indicated that some of the implementations of self-adaptation exploit more the diversity of parameter settings rather than adapting them. However, it has also been argued that the key to the success of self-adaptation seems to consist in using at the same time both a reasonably fast adaptation and reasonably large diversity to achieve a good convergence velocity and a good convergence reliability, respectively [21].

To observe the influence that the conventional/parallel application of varying mutations could have on the self-adaptive capability itself we avoid initial diversity of parameters. Experiments are conducted using populations $(\mu, \lambda) = \{(15, 100), (50, 100)\}$ and mutation ranges of $p_m = [p_m^{min}, p_m^{max}] = [1/n, \{0.50, 0.25, 0.10, 0.05\}]$. In all cases initial mutation for each individual is set to the maximum value allowed for the range, $p_m^{(t=0)} = p_m^{max}$. Fig. 1.13 **(a)** and **(b)** plot the average fitness of the best-so-far individual over the generations illustrating the convergence behavior by sGA and GA-sM, respectively. Results by GA(50,100) are also included for comparison. From these figures we can see that GA-sM achieves high fitness very rapidly in all cases. On the other hand, sGA is able to match GA-sM's convergence velocity only for small values of $p_m^{(t=0)}$. This is an indication that even in the presence of adaptation the convergence velocity of a conventional varying mutation GA would depend heavily on initial mutation rates, which is not an issue if self-adaptive mutation is applied parallel to crossover. Also, the initial lack of diversity of parameters does not affect convergence reliability of GA-sM, but it could

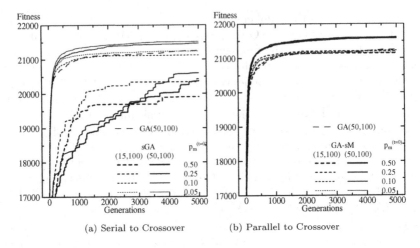

(a) Serial to Crossover   (b) Parallel to Crossover

Fig. 1.13. Self-Adaptive Varying Mutation $p_m^{(t=0)}(i) = p_m^{max}$ ($m = 30$, $n = 100$, $\phi = 0.25$)

severely affect convergence reliability of conventional sGA. Note in Fig. 1.13 **(b)** that for the same selection pressure convergence reliability of GA-sM is similar for all values of $p_m^{(t=0)}$. However, convergence reliability of sGA varies widely as shown in Fig. 1.13 **(a)**.

Similar to deterministic varying mutation, better results are achieved by $(\mu, \lambda) = (50, 100)$ rather than by $(\mu, \lambda) = (15, 100)$.

Next, we allow for initial diversity of parameters setting $p_m^{(t=0)}$ to a random value between the minimum and maximum value allowed for mutation. In this case, the disruption that higher serial mutation causes to crossover becomes less apparent due to the initial diversity of parameters and convergence speed is similar for both sGA and GA-sM. Convergence reliability of sGA also improves. However, the negative impact on reliability remains quite significant for sGA (see below). Fig. 1.14 **(a)** and **(b)** illustrates the fitness transition and the average flipped bits (Log scale) by sGA and GA-sM both with random initial mutation rates between [1/n,0.50]. Results for hGA and GA-hM are also included in Fig. 1.14 **(a)** for comparison. From these figures note that sGA converges to lower fitness and reduces mutation rates faster than GA-sM.

The self-adaptation principle tries to exploit the indirect link between favorable strategy parameters and objective function values. That is, appropriate parameters would lead to fitter individuals, which in turn are more likely to survive and hence propagate the parameter they carry with them to their offspring. A GA that applies varying mutation parallel to crossover as GA-sM can interpret better the self-adaptation principle and achieve higher performance because inappropriate mutation parameters do not disrupt crossover and it

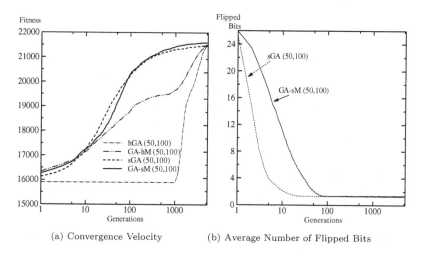

Fig. 1.14. Convergence Velocity and Average Number of Flipped Bits ($m = 30$, $n = 100$, $\phi = 0.25$). $p_m^{(t=0)} = 0.5$ for hGA and GA-hM. $p_m^{(t=0)}(i) = r$ and $[1/n, 0.5]$ for sGA and GA-SM

preserves mutation rates (see 1.5) that are being useful to the search. GA-sM self adapts mutation rates based only on the instantaneous effectiveness of varying mutations. In the case of conventional varying mutation sGA, mutation rates are self adapted based on the combined effectiveness of crossover and mutation. In other words, appropriate parameters in sGA imply mutations that would not affect greatly crossover. This introduces a selective bias towards smaller mutation rates, which can mislead the mutation rate control negatively affecting performance.

Table 1.3 and Table 1.4 also include the average percentage error gap and its standard deviation by sGA and GA-sM. The p values of the two-factor factorial ANOVA corresponding to the three experiments of reducing the feasible region (ϕ), increasing the number of constraints (m), and increasing the search space (n) are 0.0029, 0.0009, and 0.0000, respectively. These p values are considerably less than 0.05 revealing that in all three cases there is a *strong main effect* by the GA type concluding that the parallel self-adaptive varying mutation GA-sM attains significantly smaller error than the conventional self-adaptive varying mutation sGA. Furthermore, these ANOVA also revealed that GA-sM scales up better than the conventional sGA as the difficulty of the problem increases. For details see [27].

1.7 Distributed GA with Parallel Varying Mutation

In this section we study the application of parallel operators CM and SRM within a parallel distributed GA. To create a distributed GA we use a

+1+...+L communication topology [28] in which each subpopulation P_i ($i = 0, 1, ..., K - 1$) is linked to the next L subpopulations. The neighbor populations are defined by the directed links $L_{j,k}$ where

$$k = \{j + 1, ..., j + L\} \bmod K. \qquad (1.10)$$

In the above setting, the extension of GA-SRM to a distributed GA (DGA-SRM) is straightforward [29]. Extinctive selection, CM, and SRM, the basic components of the single population GA-SRM, are mostly preserved in each subpopulation $P_i(t)$ ($i = 0, 1, ..., K - 1$) at the t−th generation. CM creates offspring by conventional one-point crossover and successive background mutation operator in each $P_i(t)$. The same crossover rate p_c is used in all $P_i(t)$. The mutation probability $p_m^{(CM)}$ is set to a constant small value and is also the same in all $P_i(t)$. SRM creates offspring by an adaptive mutation operators called Adaptive Dynamic Segment (ADS) as explained in section 1.3.3. However, note that in this case the mutation segment size ℓ_i ($i = 0, 1, ..., K - 1$) is independently adjusted in each $P_i(t)$.

Migration implements a synchronous elitist broadcast strategy[30] occurring every M generations. Each subpopulation broadcasts a copy of its R best individuals to all its neighbor subpopulations. Hence, every subpopulation in every migration event receives $\lambda_m = L \times R$ migrants. In the target subpopulations, the arriving λ_m migrants replace the same number of worst performing individuals. Replacement occurs before extinctive selection. Thus, λ_m migrants also compete to survive with the best $\lambda - \lambda_m$ offspring produced by SRM and CM inside $P_i(t)$. In the following the migration rate is calculated as $100 \times \lambda_m/\lambda$. Fig. 1.16 illustrates the migration process to a given subpopulation from its two neighbors (assuming a +1+2 communication topology). As mentioned above, SRM's adaptation occurs locally in each subpopulation $P_i(t)$ but it is not realized at the generations in which migration is performed.

Fig. 1.15 illustrates a +1+2 island model in which each subpopulation is linked to two neighbors ($L = 2$). In this example, for instance, subpopulation P_0 can only send individuals to P_1 and P_2 and receive migrants from P_4 and P_5.

We test two kinds of distributed GAs in our simulations. (i) A distributed canonical GA (denoted as DGA), and (ii) the proposed distributed GA-SRM (denoted as DGA-SRM). The parameters used within each subpopulation by DGA and DGA-SRM are the same used by cGA and GA-SRM shown in Table 1.2. DGA implements the same +1+...+L communication topology and migration policy used by DGA-SRM. Unless indicated otherwise, The distributed GAs use $\lambda_{total} = 800$ individuals, $K = 16$ subpopulations ($\lambda = 50$), a 10% migration rate, and run for $T = 10^6$ function evaluations. SRM's adaptation threshold is $\tau = 0.56$. In our study we observe the influence of the problem difficulty, the subpopulation size, and the migration rate on the robustness of the distributed GAs.

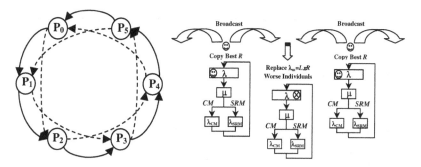

Fig. 1.15. +1+2 communication topology.

Fig. 1.16. Migration policy and extinctive selection.

1.7.1 Problem Difficulty

First we observe the performance of the distributed GAs varying the difficulty of the subclass of problems by reducing the ratio of the feasible region (ϕ), increasing the number of constraints (knapsacks m), and increasing the number of objects (size of the search space is 2^n). Results are shown in Fig. 1.17, Fig. 1.18, and Fig. 1.19, respectively. @@These and subsequent figures plot the error gap for migration intervals of $M = \{2, 5, 10, 20, 40, 100\}$ generations as well as results when no migration is used and the subpopulations evolve in total isolation (indicated by NM). As a point of reference for the quality of solutions, results obtained by the single population cGA and GA-SRM running also for $T = 10^6$ function evaluations are indicated on the left Y axis.

From these figures, we can see that DGA-SRM shows a robust and overall better performance than DGA in all subclasses of problems. Thought DGA-SRM and DGA can achieve similar or better results than its single population versions if migration is included, note that DGA-SRM even with no migration performs better than DGA with migration intervals of 5 generations. DGA can approach DGA-SRM only using very short migration intervals, i.e. every 2 generations. Different from DGA, very short migration intervals (less than 10 generation) deteriorates the performance of DGA-SRM. In short, DGA-SRM achieves high and reliable performance with less communication cost for migration, which is a big advantage for implementation.

From these figures we can also see that, similar to single population GAs, the quality of the solutions found by DGA and DGA-SRM deteriorates by increasing the difficulty of the problems either reducing the ratio of the feasible region or increasing the number of constraints.

Subpopulation Size

Second, we choose one subclass of problems ($m = 30$, $n = 100$, and $\phi = 0.25$) and observe the effect of reducing the subpopulation size while increasing

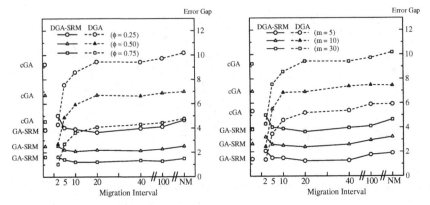

Fig. 1.17. Reducing the feasible region $\phi = \{0.75, 0.5, 0.25\}$. $K = 16$, 10% migration, $m = 30$, $n = 100$

Fig. 1.18. Increasing the number of constraints $m = \{5, 10, 30\}$. $K = 16$, 10% migration, $\phi = 0.25$, $n = 100$

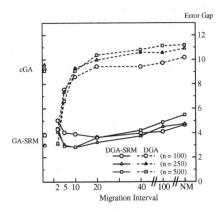

Fig. 1.19. Increasing the search space 2^n, $n = \{100, 250, 500\}$. $K = 16$, 10% migration, $\phi = 0.25$, $m = 30$

the number of subpopulations. The overall number of offspring and the total number of function evaluations are kept constant to $\lambda_{total} = 800$ individuals and $T = 10^6$ evaluations. Fig. 1.20 illustrates results by DGA and DGA-SRM using $K = \{8, 16, 32\}$ subpopulations with subpopulations sizes[2] of $\lambda = \{100, 50, 25\}$ and mutation adaptation threshold $\tau = \{0.64, 0.56, 0.50\}$, respectively. From this figure we can see that DGA-SRM tolerates population reductions better than DGA. In general, reducing the subpopulation size and increasing the number of subpopulations tends to require shorter migration

[2] When $K = 32$ DGA-SRM uses only $\lambda = 24$ to keep a 1 : 1 balance for offspring creation between CM and SRM.

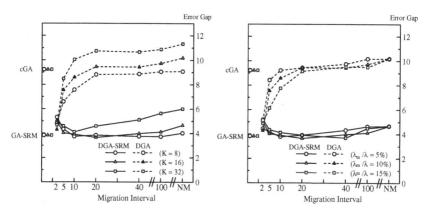

Fig. 1.20. Subpopulation size λ, K (10% migration, $m = 30$, $n = 100$, $\phi = 0.25$)

Fig. 1.21. Migration Rate λ_m/λ ($K = 16$, $m = 30$, $n = 100$, $\phi = 0.25$)

intervals to achieve high performance. Note that in the case of DGA-SRM a migration interval of $M = 100$ is enough to achieve an error similar to GA-SRM's if $K = 8$ populations of size $\lambda = 100$ are used. However, migration intervals of $M = 40$ and $M = 10$ are required for $K = 16$ ($\lambda = 50$) and $K = 32$ ($\lambda = 25$) populations, respectively. In the case of DGA, a migration interval $M = 2$ is always required to approach GA-SRM's error level.

Migration Rate

Third, the effect of the migration rate is also observed on the same subclass of problems as above ($m = 30$, $n = 100$, and $\phi = 0.25$). Fig. 1.21 illustrates results by DGA and DGA-SRM using migration rates of $\{5\%, 10\%, 15\%\}$ and $K = 16$ subpopulations. From the figure, note that in DGA-SRM smaller migration rates need shorter migration intervals and vice versa, i.e. migration intervals of $M = \{10, 20, 40\}$ work best for migration rates of $\{5\%, 10\%, 15\%\}$, respectively. To reduce communication cost, without affecting solution quality, it might be better to use larger migration intervals with higher migration rates in DGA-SRM.

Extinctive Selection

The remarkable increase in solution quality by the canonical DGA when very short migration intervals are used, i.e. 2 generations, seems at first glance rather counterintuitive since with such migrations intervals one might expect faster convergence but not higher solution quality. This is explained by the nature of the test problems and the additional selection intensity caused by migration. As mentioned before, the problems used in this study are highly

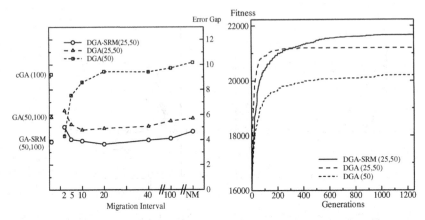

Fig. 1.22. Effect of Extinctive Selection ($K = 16$, $m = 30$, $n = 100$, $\phi = 0.25$)

Fig. 1.23. Fitness Transition ($K = 16$, $m = 30$, $n = 100$, $\phi = 0.25$, $M = 20$)

constrained with sparse feasible regions where algorithms with penalty functions have a hard time finding feasible solutions [23, 22]. A higher selection pressure in these problems is helping the algorithms to focus the search around the feasible regions. The strategies chosen in this work for migrants selection and replacement, selection of the best and replacement of the worst, cause an increase in the overall selection intensity [30]. These strategies combined with very short migration intervals are capable of producing significant selection pressures, which are being used by the DGA. In the case of DGA-SRM, the higher selection pressure is incorporated within the selection mechanism.

Fig. 1.22 illustrates the effect of extinctive selection in the distributed algorithms and clarifies the contributions of extinctive selection and parallel adaptive mutation SRM in DGA-SRM. We show results by the canonical DGA with $\lambda = 50$ individuals (DGA(50)) in each subpopulation, a DGA using (μ, λ) Proportional Selection with $\mu = 25$ parents and $\lambda = 50$ offspring in each subpopulation (DGA(25,50)), and the DGA-SRM with similar population sizes (DGA-SRM(25,50)). From this figure we see that extinctive selection alone increases the reliability of the distributed GA in this kind of problems. However, when adaptive parallel mutation SRM is used the robustness of the algorithm is improved further.

Fig. 1.23 plots the average fitness of the best solution over the generations by DGA-SRM and DGA. From this figure it can be observed that DGA-SRM has not only higher convergence reliability due to SRM but also a higher search speed caused by extinctive selection.

1.8 Summary

In this chapter we have presented a model of GA for applying crossover and varying mutation separately and in parallel. We focused on the gains on performance that can be achieved from the concurrent application of variation operators with different and complementary roles. An important conclusion of this work is that appropriate models for the concurrent application of operators, such the one proposed here, should be devised first for the actual parallelization of operators to be meaningful.

We analyzed the model using deterministic, adaptive, and self-adaptive mutation rate controls and tested its performance on a broad range of classes of 0/1 multiple knapsack problems by varying the feasible region of the search space, number of constraints, and the size of the search space. We compared the proposed model with the conventional one showing that varying mutation parallel to crossover gives an efficient framework to achieve high performance by keeping the strengths of the individual operators without interfering one with the other. We also showed that the model is superior for online adaptation of parameters and contend that it is a better option for co-adaptation of parameters.

In addition, we studied the performance of parallel operators within distributed GAs showing that the inclusion of varying mutation parallel to crossover reduces substantially communication costs due to migration, improves convergence reliability and speed of the algorithm, and can increase considerably the robustness of the algorithm. It was shown that a canonical DGA relies in higher selection intensity introduced by migration to achieve high results at the expense of very high communication cost. The distributed GA with parallel operators even without migration produced very high results compared to a canonical DGA with small migration intervals.

In the future we would like to implement this model on a parallel architecture, where in addition to the challenges imposed by the concurrent process shown in this work, we will need to cope with the demands proper of parallel implementations.

References

1. W. N. Martin, J. Lienig, and J. P. Cohoon. Island (Migration) Models: Evolutionary Algorithms Based on Punctuated Equilibria, *Handbook of Evolutionary Computation*, pp. C.6.3:1–16, Institute of Physics Publishing and Oxford University Press, New York and Bristol, 1997.
2. S. Forrest. Emergent Computation: Self-organizing, Collective, and Cooperative Phenomena in Natural and Artificial Computing Networks. *Physica*, D42:1–11, 1991.
3. V. S. Gordon and D. Whitley. Serial and Parallel Genetic Algorithms as Function Optimizers. *Proc. 5th Int'l Conf. on Genetic Algorithms*, pp. 177–183, Morgan Kaufmann, 1993.

4. S. C. Lin, W. Punch, and E. Goodman. Coarse-grain Parallel Genetic Algorithms: Categorization and New Approach. In *6th IEEE Symposium on Parallel and Distributed Processing*. IEEE Computer Society Press, 1994.

5. E. Cantú-Paz. A Survey of Parallel Genetic Algorithms. *Calculateurs Paralleles, Reseaux et Systems Repartis*, 10(2):141–171, 1998.

6. E. Alba and M. Tomassini. Parallelism and Evolutionary Algorithms. *IEEE Transactions on Evolutionary Computation*, IEEE Press, 6(5):443-462, October 2002

7. R. Hausser and R. Manner. Implementation of Standard Genetic Algorithm on MIMD Machines. In *Parallel Problem Solving from Nature III*, pp. 504–513. Springer-Verlag, 1994.

8. C. B. Pettey, M. R. Leuze, and J. J. Grefenstette. A Parallel Genetic Algorithm. *Proc. 2th Int'l Conf. on Genetic Algorithms*, pp. 155–161, Lawrence Erlbaum Assoc., 1987.

9. R. Tanese. Parallel Genetic Algorithms for a Hypercube. *Proc. 2th Int'l Conf. on Genetic Algorithms*, pp. 177–183, Lawrence Erlbaum Assoc., 1987.

10. B. Manderick and P. Spiessens. Fine-grained Parallel Genetic Algorithms. *Proc. 3th Int'l Conf. on Genetic Algorithms*, pp. 428–433, Morgan Kaufmann, 1989.

11. H. Muhlenbein. Parallel Genetic Algorithms, Population Genetics and Combinatorial Optimization. *Proc. 3th Int'l Conf. on Genetic Algorithms*, pp. 416–421, Morgan Kaufmann, 1989.

12. John H. Holland. *Adaptation in Natural and Artificial Systems*. University of Michigan Press, 1975.

13. D. E. Goldberg. *Genetic Algorithms in Search, Optimization and Machine Learning*. Reading. Addison-Wesley, 1989.

14. A. E. Eiben, R. Hinterding, and Z. Michalewicz. Parameter Control in Evolutionary Algorithms. *IEEE Transactions on Evolutionary Algorithms*, 3(2):121–141, July 1996.

15. H. Aguirre, K. Tanaka, and T. Sugimura. Cooperative Model for Genetic Operators to Improve GAs. In *Proc. IEEE Int'l Conf. on Information, Intelligence, and Systems*, pp. 98–106, 1999.

16. H. Aguirre, K. Tanaka, T. Sugimura, and S. Oshita. Cooperative-Competitive Model for Genetic Operators: Contributions of Extinctive Selection and Parallel Genetic Operators. In *Proc. Late Breaking Papers Genetic and Evolutionary Computation Conference*, pp. 6–14. Morgan Kaufmann, 2000.

17. H. Aguirre, K. Tanaka, and T. Sugimura. Empirical Model with Cooperative-Competitive Genetic Operators to Improve GAs: Performance Investigation with 0/1 Multiple Knapsack Problems IPSJ Journal, vol. 41, no. 10, pp. 2837-2851, 2000.

18. W. Spears. *Evolutionary Algorithms: The Role of Mutation and Recombination*. Springer-Verlag, 2000.

19. T. Bäck. *Evolutionary Algorithms in Theory and Practice*. Oxford University Press, 1996.

20. T. Bäck and M. Schutz. Intelligent Mutation Rate Control in Canonical Genetic Algorithms. In *Lecture Notes on Artificial Intelligence*, volume 1079, pp. 158–167. Springer, 1996.

21. Thomas Bäck. Self-adaptation, *Handbook of Evolutionary Computation*, pp. C7.1:1–13, Institute of Physics Publishing and Oxford University Press, New York and Bristol, 1997.

22. Z. Michalewicz. *Genetic Algorithms + Data Structures = Evolution Programs*. Springer-Verlag, third revised and extended edition, 1996.
23. P.C. Chu and J. E. Beasley. A Genetic Algorithm for the Multidimensional Knapsack Problem. *Journal of Heuristics*, 4:63–86, 1998.
24. M. Shinkai, H. Aguirre, and K. Tanaka. Mutation Strategy Improves GA's Performance on Epistatic Problems. In *Proc. 2002 IEEE World Congress on Computational Intelligence*, pp. 795–800, 2002.
25. H. Aguirre and K. Tanaka. Parallel Varying Mutation Genetic Algorithms. In *Proc. 2002 IEEE World Congress on Computational Intelligence*, pp. 795–800, 2002.
26. H. Aguirre and K. Tanaka. Modeling Efficient Parallel Varying Mutation Genetic Algorithms. In *Proc. Workshop Program 2002 Genetic and Evolutionary Computation Conference*, pp. 256–259, 2002.
27. H. Aguirre and K. Tanaka. Parallel Varying Mutation in Deterministic and Self-adaptive GAs. In *Proc. Seventh International Conference on Parallel Problem Solving from Nature (PPSN 2002)*, volume 2439 of *Lecture Notes in Computer Science*, pp. 111–121. Springer-Verlag, 2002.
28. E. Cantú-Paz. Topologies, Migration Rates, and Multi-population Parallel Genetic Algorithms. *Proc. Genetic and Evolutionary Computation Conference*, pp. 91–98, Morgan Kaufmann, 1999.
29. H. Aguirre, K. Tanaka, T. Sugimura, and S. Oshita. Increasing the Robustness of Distributed Genetic Algorithms by Parallel Cooperative-Competitive Genetic Operators. *Proc. Genetic and Evolutionary Computation Conference*, pp. 195–202, Morgan Kaufmann, 2001.
30. E. Cantú-Paz. Migration Policies, Selection Pressure, and Parallel Evolutionary Algorithms. *Proc. Late Breaking Papers Genetic and Evolutionary Computation Conference*, pp. 65–73. Morgan Kaufmann, 1999.

2

Parallel Evolutionary Multiobjective Optimization

Francisco Luna, Antonio J. Nebro, and Enrique Alba

Dpto. Lenguajes y Ciencias de la Computación, E.T.S.I. Informática
University of Málaga, Campus de Teatinos 29071 Málaga, Spain
{antonio,flv,eat}@lcc.uma.es

Research on multiobjective optimization is very active currently because most of the real-world engineering optimization problems are multiobjective in nature. Multiobjective optimization does not restrict to find a unique single solution, but a set of solutions collectively known as the Pareto front. Evolutionary algorithms (EAs) are especially well-suited for solving such kind of problems because they are able to find multiple trade-off solutions in a single run. However, these algorithms may be computationally expensive because (1) real-world problem optimization typically involves tasks demanding high computational resources and (2) they are aimed at finding the whole front of optimal solutions instead of searching for a single optimum. Parallelizing EAs arises as a possible way of facing this drawback. The first goal of this chapter is to provide the reader with a wide overview of the literature on parallel EAs for multiobjective optimization. Later, we include an experimental study where we develop and analyze pPAES, a parallel EA for multiobjective optimization based on the Pareto Archived Evolution Strategy (PAES). The obtained results show that pPAES is a promising option for solving multiobjective optimization problems.

2.1 Introduction

In the last years, much attention has been paid to the optimization of problems whose formulation involves optimizing more than one objective function [16, 23, 72], being this interest mainly motivated by the multiobjective nature of most real-world problems. The task of finding solutions for such kind of problems is known as *multiobjective optimization* (MO). Multiobjective optimization problems (MOPs) have therefore a number of objective functions to be optimized which are usually in conflict with each other.

F. Luna et al.: *Parallel Evolutionary Multiobjective Optimization*, Studies in Computational Intelligence (SCI) **22**, 33–56 (2006)
www.springerlink.com

Generally speaking, multiobjective optimization does not restrict to find a unique single solution, but a set of solutions called *nondominated solutions*. Each solution in this set is said to be a *Pareto optimum*, and when they are plotted in the objective space they are collectively known as the *Pareto front*. Obtaining the Pareto front of a given MOP is the main goal of multiobjective optimization.

In order to search for such set of Pareto optima, Evolutionary Algorithms (EAs) are especially well-suited because of their ability for finding multiple trade-off solutions in one single run. Well-accepted subclasses of EAs are Genetic Algorithms (GA), Genetic Programming (GP), Evolutionary Programming (EP), and Evolution Strategies (ES). Multiobjective EAs (MOEAs) have been investigated by many authors, and some of the most well-known algorithms for solving MOPs, such as NSGA-II [24], PAES [55], MOGA [33], microGA [15], and SPEA2 [97], belong to this kind of techniques.

However, MOEA approaches often need to explore larger portions of the search space because they require the whole Pareto front to be obtained, i.e., not a single optimum but a set of Pareto optima, thus resulting in a huge number of function evaluations. Additionally, many real-world MOPs typically use computationally expensive methods for computing the objective functions and constraints. These two facts hinder MOEAs from solving efficiently such real-world applications.

These drawbacks can be faced in two possible ways. On the one hand, surrogate models of the fitness functions can be used instead of true fitness function evaluation [36]. On the other hand, MOEAs can be parallelized so that evaluations are performed on multiple processors. The last issue will be addressed in this chapter.

Due to their population-based approach, EAs are very suitable for parallelization (see Figure 2.1) because their main operations (i.e., crossover, mutation, and in particular function evaluation) can be carried out independently on different individuals. There is a vast literature on how to parallelize EAs (the reader is referred to [6, 7, 12] for surveys on this topic). However, parallelism is not only a way for solving problems more rapidly, but for developing new and more efficient models of search: a parallel EA can be more

```
t := 0;
Initialization:  P(0) := {a_1(0),..., a_μ(0)} ∈ I^μ;
Evaluation:  P(0) : {Φ(a_1(0)),..., Φ(a_μ(0))};
while ι(P(t)) ≠ true do
        Selection:  P'(t) := s_Θ_s (P(t));
        Crossover:  P''(t) := ⊗_Θ_c (P'(t));
        Mutation:  P'''(t) := m_Θ_m (P''(t));
        Evaluation:  P'''(t) : {Φ(a'''_1(t)),..., Φ(a'''_λ(t))};
        Replacement:  P(t + 1) := r_Θ_r (P'''(t) ∪ Q);
        <Communication>
        t := t + 1;
end while
```

Fig. 2.1. Pseudocode of a parallel EA

effective than a sequential one, even when executed on a single processor. The advantages that parallelism offers to single objective optimization should hold as well in multiobjective optimization, but few efforts have been devoted to parallel implementations in this field until recently [16].

The first contribution of this chapter is to provide the reader with a wide review of works related to parallel MOEAs. We have focussed mainly on those works dealing with physical parallelizations. We have followed the traditional categorization of parallel EAs [6] as a starting point. Second, we explore the use of parallelism by developing a new MOEA, which is called pPAES, a parallel search model based on PAES [55], a usual standard in MOEAs. The pPAES algorithm has been tested on two mathematical test functions (Fonseca [34] and Kursawe [56]) and a real-world MOP coming from the telecommunication industry, the CRND problem [65].

The remainder of the chapter is structured as follows. In the next section, we present a brief theoretical background on multiobjective optimization as well as a survey of parallel EAs for multiobjective optimization. Section 2.3 is aimed at introducing the pPAES algorithm, the experimentation carried out, and the analysis of the results. Finally, we summarize the conclusions and discuss several lines for future research in Section 2.4.

2.2 Multiobjective Optimization

In this section, we start explaining some basic multiobjective optimization concepts. Next, we review some related works about using parallel MOEAs. We also present other non-MOEA parallel algorithms for solving MOPs at the end.

2.2.1 Theoretical Background

Let us begin with the multiobjective optimization theoretical background. A general multiobjective optimization problem (it is assumed, without loss of generality, minimization problems) can be formally defined as follows:

Definition 1 (MOP). *Find a vector $\boldsymbol{x}^* = [x_1^*, x_2^*, \ldots, x_n^*]$ which satisfies the m inequality constraints $g_i(\boldsymbol{x}) \geq 0, i = 1, 2, \ldots, m$, the p equality constraints $h_i(\boldsymbol{x}) = 0$, $i = 1, 2, \ldots, p$, and minimizes the vector function $\boldsymbol{f}(\boldsymbol{x}) = [f_1(\boldsymbol{x}), f_2(\boldsymbol{x}), \ldots, f_k(\boldsymbol{x})]^T$, where $\boldsymbol{x} = [x_1, x_2, \ldots, x_n]^T$ is the vector of decision variables.* ∎

The set of all values satisfying the constraints defines the *feasible region* Ω and any point $\boldsymbol{x} \in \Omega$ is a *feasible solution*. As mentioned before, we seek for the *Pareto optimum*. Its formal definition is provided next:

Definition 2 (Pareto Optimality). *A point $\boldsymbol{x}^* \in \Omega$ is Pareto Optimal if for every $\boldsymbol{x} \in \Omega$ and $I = \{1, 2, \ldots, k\}$ either, $\forall_{i \in I}(f_i(\boldsymbol{x}) = f_i(\boldsymbol{x}^*))$ or, there is at least one $i \in I$ such that $f_i(\boldsymbol{x}) > f_i(\boldsymbol{x}^*)$.* ∎

This definition states that $\boldsymbol{x}^*$ is Pareto optimal if there exists no feasible vector $\boldsymbol{x}$ which would decrease some criterion without causing a simultaneous increase in at least one other criterion. Other important definitions associated with Pareto optimality are the following:

Definition 3 (Pareto Dominance). *A vector $\boldsymbol{u} = (u_1, \ldots, u_k)$ is said to dominate $\boldsymbol{v} = (v_1, \ldots, v_k)$ (denoted by $\boldsymbol{u} \preccurlyeq \boldsymbol{v}$) if and only if u is partially less than v, i.e., $\forall i \in \{1, \ldots, k\}, \; u_i \leq v_i \wedge \exists i \in \{1, \ldots, k\} : u_i < v_i$.* ∎

Definition 4 (Pareto Optimal Set). *For a given MOP $\boldsymbol{f}(\boldsymbol{x})$, the Pareto optimal set is defined as $\mathcal{P}^* = \{\boldsymbol{x} \in \Omega | \neg \exists \boldsymbol{x}' \in \Omega, \boldsymbol{f}(\boldsymbol{x}') \preccurlyeq \boldsymbol{f}(\boldsymbol{x})\}$.* ∎

Definition 5 (Pareto Front). *For a given MOP $\boldsymbol{f}(\boldsymbol{x})$ and its Pareto optimal set $\mathcal{P}^*$, the Pareto front is defined as $\mathcal{PF}^* = \{\boldsymbol{f}(\boldsymbol{x}), \boldsymbol{x} \in \mathcal{P}^*\}$.* ∎

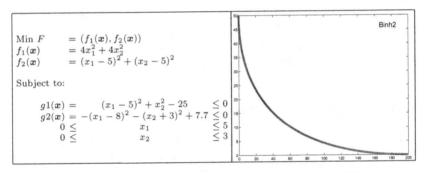

Fig. 2.2. Formulation and Pareto front of the problem Bihn2

That is, the Pareto front is composed of the values in the objective space of the Pareto optimal set. As an example, in Figure 2.2 we show the formulation and the Pareto front of a constrained MOP called Bihn2 [16]. It is a two-objective problem with a single convex Pareto front.

2.2.2 A Survey of Parallel MOEAs

Parallelism arises naturally when dealing with populations of individuals, since each individual is an independent unit. As a consequence, the performance of population-based algorithms is specially improved when run in parallel. Two parallelizing strategies are specially relevant for population-based algorithms: (1) parallelization of computation, in which the operations commonly applied to each individual are performed in parallel, and (2) parallelization of population, in which the population is split in different parts, each one evolving in semi-isolation (individuals can be exchanged between subpopulations).

The simplest parallelization scheme of this kind of algorithms is the well-known *Master-Slave* or *global parallelization* method (Figure 2.3a). In this

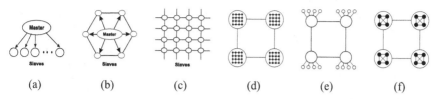

(a)   (b)   (c)   (d)   (e)   (f)

Fig. 2.3. Different models of parallel EAs: (a) global parallelization, (b) coarse grain, and (c) fine grain. Many hybrids have been defined by combining parallel EAs at two levels: (d) coarse and fine grain, (e) coarse grain and global parallelization and (f) coarse grain at the two levels

scheme, a central processor performs the selection operations while the associated slave processors perform the recombination, mutation, and/or the evaluation of the fitness function. This algorithm is the same as the sequential one, although it is faster, especially for time-consuming objective functions.

However, many parallel EAs found in the literature utilize some kind of spatial disposition of the individuals (it is said that the population is then structured), and afterwards parallelize the resulting chunks in a pool of processors. Among the most widely known types of structured EAs, the *distributed* (dEA) (or coarse-grain) and *cellular* (cEA) (or fine-grain) algorithms are very popular optimization procedures [6]. In the case of distributed EAs (Figure 2.3b), the population is partitioned in a set of islands in which isolated EAs run in parallel. Sparse individual exchanges are performed among these islands with the goal of inserting some diversity into the subpopulations, thus avoiding them to fall into local optima. In the case of a cellular EA (Figure 2.3c), subpopulations are typically composed of 1 or 2 individuals, which may only interact with its nearby neighbors in the breeding loop, i.e., the concept of *neighborhood* is introduced. This parallel scheme is targeted to massively parallel computers. Also, hybrid models have been proposed (Figure 2.3d–f) in which a two-level approach of parallelization is undertaken. In these models, the higher level for parallelization uses to be a coarse-grain implementation and the basic island performs a cEA, a master-slave method, or even another distributed one.

This taxonomy holds as well for parallel MOEAs, so we can consider Master-Slave MOEAS (*msMOEAs*), distributed MOEAS (*dMOEAs*), and cellular MOEAs (*cMOEAs*). Nevertheless, these two decentralized population approaches need a further particularization. As we stated before, the main goal of any multiobjective optimization algorithm is to hopefully find the optimal Pareto front for a given MOP. It is clear that in msMOEAs the management of this Pareto front is carried out by the master processor. But, when the search process is distributed among different subalgorithms, as it happens in dMOEAs and cMOEAs, the management of the nondominated set of solutions during the optimization procedure becomes a capital issue. Hence, we distinguish whether the Pareto front is distributed and locally managed by each subEA during the computation, or it is a centralized element of the algorithm. We call them *Centralized Pareto Front* (CPF) structured MOEAs and *Distributed Pareto Front* (DPF) structured MOEAs, respectively. We want to remark that this taxonomy does make sense when Pareto optimality is addressed since other approaches can be taken for dealing with multiple objective functions [16, 23].

Among the revised literature analyzed for this chapter, no fully CPF implementations of decentralized-population MOEAs have been found, mainly motivated by efficiency issues that researchers typically have to address. All the CPF decentralized MOEAs in the literature are combined with DPF phases where local nondominated solutions are considered. After each DPF phase, a single optimal Pareto front is built by using these local Pareto optima. Then, the new Pareto front is again distributed for local computation, and so on.

In Table 2.1 we show a list of works (sorted by year) on parallel MOEAs which have been actually deployed on parallel platforms. The following information is presented in the table (in order of appearance):

- the reference(s) of the work,
- the year of the first publication related to the algorithm,
- the parallelization strategy used (PS),
- whether the algorithm considers or not Explicit Pareto Optimality (EPO),
- the membership of dMOEAs or cMOEAs to the two subcategories defined previously, CPF or DPF (when applicable),
- the implementation language and/or communication library used,
- the communication topology of the parallel MOEA,
- a brief description of the algorithm, and
- the application domain.

Figure 2.4 displays the number of publications on parallel MOEAs in the literature through the years. Given the inherent parallelism of MOEAs as well as the current cheap access to parallel platforms (especially cluster of workstations), it is somewhat surprising that the interest in this field is not yet evident. As Figure 2.4 shows, only 6 works on parallel MOEAs were published in 2004 (this chapter has been written at the beginning of 2005 and this explains the lack of references this year).

The number of publications on each parallel scheme of MOEAs is shown in Figure 2.5. It can be seen that the most popular MOEA parallelizations are msMOEA and dMOEA. The msMOEA scheme is commonly used because, on the one hand, computationally expensive objective function evaluations are usually involved in many real-world MOPs and, on the other hand, they admit a direct parallelization scheme based on sequential MOEAs. In the case of dMOEAs, the intensive research on their single objective counterparts (dEAs) has led their multiobjective extension to be widely used. The shortage of publications concerning the cMOEA parallel scheme is inherited from the diffusion model itself (cEA), which was originally targeted to massively

Table 2.1. Parallel MOEAs

References	Year	PS	EPO	CPF	DPF	Language	Topology	Description	Application Domain
[59]	1993	dMOEA	●	–	–	C	Fully Connected	DCGA: one island per objective	Fuzzy control problem
[83]	1995	msMOEA	●	–	–	–	Master/Slave	GA running on the Internet	Microprocessor design
[29, 28, 30]	1996	dMOEA	√	●	–	MPI	Torus	dGA with supervised evolution	Aerodynamic design
[60]	1996	msMOEA	√	●	√	MPI	Master/Slave	Global parallelization of a GA	Airfoil design
[79]	1996	cMOEA	√	●	–	–	Torus	Parallel cellular GA	Sensibility analysis problem
[58]	1997	dMOEA	√	●	–	C++	Torus	dGA running on a cluster	VLSI routing
[68, 69, 67]	1997	msMOEA	√	●	√	–	Master/Slave	Global parallelization of MOGA [33]	Airfoil design
[48]	1998	msMOEA	√	●	√	MPI	Master/Slave	Parallel GA with global parallelization	Airfoil design
[76, 88]	1998	dMOEA	√	●	√	C?	Master/Slave	GA for parallel shared memory machines	Wing design
[10, 18]	1999	dMOEA	√	●	√	–	Fully Connected	Decision-making unit rule-based GA	Control system design
[42, 43, 90]	1999	dMOEA	√	●	√	MPI	Star	Divided Range MOGA (DRMOGA)	Mathematical test functions
[38, 37]	2000	msMOEA	√	●	●	PVM	Master/Slave	Global parallelization of the NPGA [45]	Astrophysical problem
[75]	2000	msMOEA	√	●	●	CORBA	Master/Slave	Global parallelization of a MOGA	Yacht fin keel design
[63]	2000	msMOEA	●	●	–	MPI	Master/Slave	GA with parallel genetic operators	Shape design
[65]	2000	msMOEA	√	●	√	C/PVM	Master/Slave	Global parallelization of a GA	Radio network design
[78]	2000	dMOEA	●	●	√	–	Star	Islands optimizing different objectives	Actuators placement
[31]	2001	msMOEA	√	●	●	C/PVM	Fully Connected	Islands separately executing SPEA	Network design
[81]	2001	msMOEA	●	●	√	–	Master/Slave	GA parallelized on a SGI ORIGIN 2000	Wing design
[91]	2001	msMOEA	●	●	√	MPI?	Master/Slave	Local Cultivation model GA	Antenna placement
[25, 26]	2002	dMOEA	√	√	√	C/MPI	Fully Connected	Island model of NSGA-II	Mathematical test functions
[20, 21, 22]	2002	dMOEA	√	√	√	C/MPI	Star	Islands exploring different regions	Mathematical test functions
[44]	2002	dMOEA	√	●	–	–	Dynamic	DGA with asynchronous migrations	Mathematical test functions
[50]	2002	dMOEA	√	√	–	–	Fixed Graph	Islands with different weights	Mathematical test functions
[51]	2002	msMOEA	√	√	√	C/MPI	Master/Slave	Global parallelization of DRMOGA [42]	Car engine design
[52]	2002	dMOEA	√	●	√	PVM	Fully Connected	MOGA + coevolutionary GA	Mathematical test functions
[89]	2002	dMOEA	●	●	●	C/Globus	Hierarchy	dMOEA based on Nash equilibrium	Aerodynamic design
[8, 19]	2003	msMOEA	●	●	√	PVM	Master/Slave	MicroGA parallelization using Globus	Diesel engine chamber design
[32]	2003	msMOEA	●	●	●	C++/MPI	Master/Slave	Parallel ES on a cluster	Multibody system optimization
[54]	2003	msMOEA	●	●	●	C++/MPI	Star	Farming model of a MOEA	Groundwater bioremediation
[62]	2003	msMOEA	√	√	√	Java	Master/Slave	Global parallelization of a GA	Optical filter design
[71, 77]	2003	msMOEA	√	√	√	Java	Master/Slave	Global parallelization of NSGA	Handwritten recognition
[86]	2003	dMOEA	√	●	√	Java	Directed Graph	dMOEA P2P parallel implementation	Mathematical test functions
[93]	2003	dMOEA	√	√	√	–	dMOEA	Specialized Island Model	Mathematical test functions
[94]	2003	dMOEA	√	√	√	C/MPI	Fully Connected	Parallel SPEA2 for SMP machines	Mathematical test functions
[11]	2004	dMOEA	√	√	√	C/MPI	dMOEA	Cone separation approach of NSGA-II	Mathematical test functions
[14]	2004	dMOEA	√	√	√	MPI	Master/Slave	Parallel coevolutionary EA	Quadratic assignment problem
[53]	2004	dMOEA	√	√	√	–	–	Island model of MOMGA-II	Diesel engine design
[40, 41]	2004	msMOEA	√	√	√	MPI	Master/Slave	Parallel neighborhood cultivation GA	De Novo peptide identification
[61]	2004	dMOEA	√	●	√	MPI	Random Ring	Parallel function evaluation dMOEA	Autonomous controller design
[70]	2004	msMOEA	√	●	√	MPI	Master/Slave	Parallel multiobjective GP	Mathematical test functions
[64]	2005	dMOEA	√	√	√	MPI	Hypergraph	Island model of NSGA-II	Mathematical test functions
[85]	2005	dMOEA	√	√	√	–	Star	MOEA using clustering algorithms	Mathematical test functions
[96]	2005	dMOEA	√	●	√	C?	Fully Connected	Island model of SPEA	Mathematical test functions

√ = Yes, ● = No, – = Not applicable/Not available

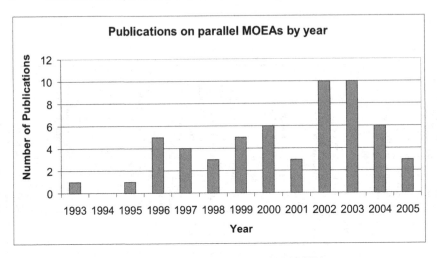

Fig. 2.4. Publications on parallel MOEAs

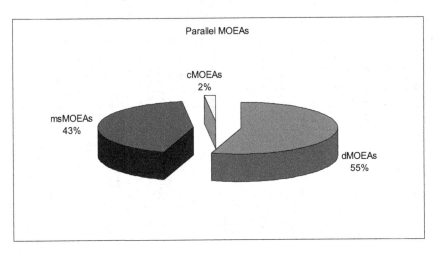

Fig. 2.5. Number of publications for each type of parallel MOEAs

parallel machines. Since this kind of machines fell into disuse, cellular models are usually executed on sequential machines at present.

We have included the EPO column in Table 2.1 because other MOP-like methods have been considered in the literature, being the aggregative weighted sum the most popular technique [50, 83]. However, it can also be found specific approaches like using trade-off analyses via Nash equilibrium [89]. Note that when a parallel MOEA does not address Pareto optimality (see "•" marks of EPO column in Table 2.1), the further categorization for dMOEAs and cMOEAs does not make sense ("–" marks in columns CPF and DPF) because there not exist Pareto fronts to be considered. Nevertheless, most recent parallel MOEAs are based on the Pareto optimality approach, as shown in Table 2.1, and MO researchers even classify weighted sums as non MO algorithms.

The "Language" column in Table 2.1 shows that C/C++ are the preferred programming languages for this kind of algorithms, mainly because of their efficiency. FORTRAN [75] and Java [94] implementations have also been addressed. They are typically combined with communication libraries such as PVM [35] or MPI [39] in order to develop parallel programs. Besides, works using modern Grid technologies to parallelize MOEAs [8] are available.

Concerning the topology of parallel MOEAs, it is obvious that msMOEAs use a *Master/Slave* topology. In the decentralized models, *Star* topologies are widely used for gathering operations to create a single Pareto front. In them, a central node collects nondominated solutions from the other nodes, performs some operations on the resulting front, and then sends the resulting Pareto optimal solutions back to the nodes. Table 2.1 also shows that *Fully Connected* topologies are very frequent. In this approach, one or more nondominated solutions are periodically broadcast to the subpopulations to enhance the diversity of the search.

A wide spectrum of applications can be observed in the last column of Table 2.1. As stated before, most of them are real-world problems. In fact, they are mainly engineering design problems: VLSI design, aerodynamic airfoil design, or radio network design. Another important MOPs are mathematical test functions [16, 87] which are used for evaluating and comparing new multiobjective algorithms. Figure 2.6 gives us insights on the importance of these two kinds of MOPs. It can be seen that they cover more than 80% of works on parallel MOEAs.

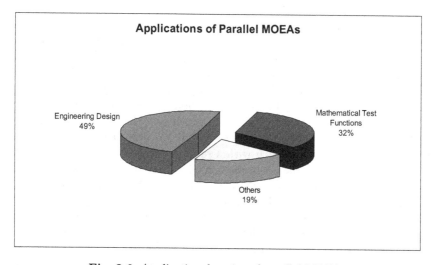

Fig. 2.6. Application domains of parallel MOEAs

Table 2.2. Parallel algorithms for solving MOPs

References	Year	Description
[17]	1994	Parallel simulated annealing for VLSI optimization
[92]	1994	Parallel exact algorithm for solving multiple objective linear programs
[13]	1998	Parallel cooperative simulated annealing threads for SVC planning
[2, 3, 4, 80]	2000	Parallel Tabu Search for VLSI design
[1]	2000	Parallel simulated annealing for combinatorial library design
[27]	2000	Ant colony optimization for scheduling problems
[84]	2003	Exact technique using a SQP algorithm for analog circuit analysis
[46, 47]	2004	Parallel tabu search for aerodynamic shape optimization
[57]	2004	Parallel exact method to solve bicriteria flow-shop problems
[66]	2004	Distributed enumerative search algorithm using grid computing
[74]	2004	Vector evaluated particle swarm optimization
[73]	2004	Vector evaluated differential evolution
[95]	2004	Adaptive Pareto differential evolution

2.2.3 Other Parallel Algorithms for Solving MOPs

Although MOEAs are the most widely used approaches for MOPs due to their inherent parallelism, parallel algorithms based on exact or other metaheuristic techniques can be found in the literature as well. We summarize them in Table 2.2, where we include the reference(s) of the work, the year of the first publication related to the algorithm, and a brief description. Among the exact techniques, an enumerative search algorithm based on Condor to distribute the computation is presented in [66]. In this work, the search space of various mathematical test MOPs is divided into several subspaces, each one explored by a sequential enumerative algorithm. Computing exact Pareto fronts is extremely useful for the community and is a time consuming task only solvable by exact methods. Hybrid strategies with different metaheuristics [49] and exact and heuristic methods [9] have also been studied.

As to other parallel non-MOEA metaheuristics, a Particle Swarm Optimization (PSO) algorithm called VEPSO has been presented in [73]. VEPSO, which is based on VEGA [82], uses several collaborative swarms to explore the search space. Each swarm is evaluated with a single objective function, but information coming from other swarm(s) is used for influencing its motion in the search space. VEPSO has been parallelized with PVM by distributing each swarm on a different processor. Also, Zaharie and Petcu have proposed the parallel Adaptive Pareto Differential Evolution (APDE) in [95], a population-based algorithm for continuous domains which uses the ranking method and crowding procedures presented in [24]. APDE has been parallelized by using a multiple population approach with a random topology for migration operations. There exist parallel Tabu Search (TS) algorithms for solving MOPs as well. In [46], a Master/Slave parallelization of a TS for aerodynamic shape optimization is presented. While the main multiobjective TS method runs on the master program, the slaves perform the objective function evaluation in parallel, so multiple sampling points in the neighborhood are explored at each step. Agrafiotis has presented in [1] a parallel Simulated Annealing algorithm for the multiobjective design of combinatorial libraries. The algorithm allows different threads to follow their own independent Monte Carlo trajectories during each temperature cycle in parallel. Threads synchronize at the end of each cycle, and the best among the last states visited by each thread is recorded and used as the starting point for the next iteration.

2.3 A Parallel MOEA: pPAES

To offer the reader an example algorithm we include an experimental study in this section. For this goal we have developed a parallel MOEA, named pPAES, which is based on the Pareto Archived Evolution Strategy (PAES) [55]. PAES is a well-known sequential multiobjective algorithm against which new proposals are compared.

The basic PAES version used here is a $(1 + 1)$ evolution strategy (Section 2.3.1) employing local search and a reference archive of previously found solutions in order to identify the approximate dominance ranking of the current and candidate solutions vectors. The pPAES algorithm is a parallel extension consisting of several islands executing the sequential PAES and exchanging solutions among them. It is aimed at, first, reducing the execution times of PAES and, second, enhancing its search model to hopefully get higher quality solutions. We will detail pPAES later in Section 2.3.2.

2.3.1 $(1 + 1)$-PAES

We have implemented a C++ version of the $(1 + 1)$-PAES algorithm based on the description presented in [55]. The $(1 + 1)$-PAES represents the simplest nontrivial approach to a multiobjective local search procedure and will be used here for benchmark our pPAES proposal. An outline of this algorithm is presented in Fig. 2.7. PAES is based on maintaining a single solution that is mutated to generate a new candidate solution at each iteration (line 3 in Fig. 2.7). Afterwards, the algorithm determines whether to accept or to reject

```
1 generate initial random solution c and add it to the archive
2 while termination criterion does not hold
3 mutate c to produce m and evaluate m
4 if (c dominates m)
5 discard m
6 else if (m dominates c)
7 replace c with m, and add m to the archive
8 else if (m is dominated by any member of the archive)
9 discard m
10 else
11 apply test(c,m,archive) to determine the new current solution
and whether to add m to the archive
12 end while
```

Fig. 2.7. Pseudocode for $(1 + 1)$-PAES

the mutant solution and whether to archive or not it in a list of nondominated solutions by means of an acceptance criterion. Since we intend to solve real-world problems whose evaluation requires a high computational cost, a basic improvement has been performed with respect the traditional PAES: if the mutation operator does not modify the current individual, c, then neither the evaluation nor the archiving procedure are done. In this case, the termination criterion is to reach a preprogrammed number of function evaluations.

Since the aim of the multiobjective search is to find a set of equally spread nondominated solutions, PAES uses a crowding procedure based on an adaptive numerical grid that recursively divides up the objective space [55]. When each solution is generated, its *grid location* in the objective space is determined by a recursive subdivision. The number of solutions currently residing in each grid location is also maintained in order to make decisions in the selection and in the archiving procedure when a location is too crowded.

2.3.2 pPAES

The pPAES algorithm is a dMOEA which uses a DPF strategy where each island executes a $(1 + 1)$-PAES; pPAES works as follows. Each process computes PAES during a predefined number of function evaluations and maintains its own local archive of nondominated solutions. This Pareto optimal set has the same maximum size, N, in all the islands.

Periodically, a synchronous migration operation produces an exchange of solutions in a unidirectional ring topology. This migration process requires further explanations. First, we must set how many and what individuals should be migrated. The pPAES algorithm migrates one single individual each time, which is uniformly randomly chosen from the local archive of the island. Then, the new individual is included as a normal mutated individual m in the local PAES (line 3 in Fig. 2.7). Second, the frequency of the migration operation must be defined. Exchanging solutions in pPAES is carried out after a fixed step number, called *migration frequency*, which is measured in terms of the number of function evaluations. The value for this parameter has been set to 30.

The last step in pPAES consists in building the Pareto front that will be presented as the final result. This front will have the same number of non-dominated solutions as a single PAES algorithm. So, if pPAES uses p parallel processes, then, at most, $p \cdot N$ Pareto optima could have been calculated during the optimization procedure, because it could happen that not all the islands calculate the maximum number of N solutions. This way, pPAES gathers all this information to present a single front. All pPAES processes send their nondominated solutions to a distinguished process that includes them (along with its already locally stored solutions) in a single front by using the adaptive grid algorithm presented in the previous section.

Table 2.3. Formulation of the multiobjective problems Fonseca and Kursawe

Problem	Definition	Constraints
	$\text{Min } F = (f_1(x), f_2(x))$	
Fonseca	$f_1(x) = 1 - e^{-\sum_{i=1}^{n}(x_i - \frac{1}{\sqrt{n}})^2}$	$-4 \leq x_i \leq 4$
	$f_2(x) = 1 - e^{-\sum_{i=1}^{n}(x_i + \frac{1}{\sqrt{n}})^2}$	$i \quad = 1, 2, 3$
	$\text{Min } F = (f_1(x), f_2(x))$	$-5 \leq x_i \leq 5$
Kursawe	$f_1(x) = \sum_{i=1}^{n-1}(-10e^{(-0.2*\sqrt{x_i^2 + x_{i+1}^2})})$	$i \quad = 1, 2, 3$
	$f_2(x) = \sum_{i=1}^{n}(|x_i|^a + 5\sin(x_i)^b)$	$a \quad = 0.8$
		$b \quad = 3$

2.3.3 Experiments

Here, we first present the benchmark used for evaluating pPAES, which is composed of two mathematical multiobjective functions and a real-world, complex, time-consuming MOP coming from the telecommunication industry. Next, the performance metrics used for measuring the results are described. Finally, we analyze the obtained results.

All the pPAES experiments have been deployed on a cluster of 16 PCs equipped with Pentium 4 processors at 2.8 GHz, 512 MB of RAM, and being the interconnection network a Fast-Ethernet at 100 Mbps. All the programs have been compiled with GNU gcc v3.2 using the option -O3. The parallel versions use MPICH 1.2.5, an implementation of the standard MPI.

Benchmark

Two traditional mathematical MOPs selected from the especialized literature have been used for testing the search model proposed by pPAES: Fonseca [34] and Kursawe [56]. Their definition is shown in Table 2.3.

The pPAES algorithm has been also used for solving a cellular radio network design (CRND) problem coming from the telecommunication industry [65]. In this chapter, the two functions are used for showing the working principles of the algorithm while the CRND problem is used for illustrating the actual power of our approach.

The CRND problem may be reduced to the placement and the configuration of base stations (BSs) on candidate sites. The main decision variable of the problem deals with the mapping of BSs on the potential sites. BSs are of three types: omnidirectional, small directive, or large directive. Each site may be equipped with either one BS with a single omnidirectional antenna or with one to three BS, each having a directive antenna. In addition to the mapping decision variable and the type of antenna, each BS is configured by some engineering parameters: the azimuth, that is the direction the BS is pointing to, the transmitter power, and the vertical tilt.

We define three main objectives for the problem: minimize the number of sites used (reducing the number of sites reduces the cost of the design), maximize the amount of traffic held by the network, and minimize the interferences. Two main constraints have to be satisfied to design a cellular network:

- Coverage of the area. All the geographical area must be covered with a minimal radio field value, that must be greater than the receiver sensibility threshold of the mobile device.
- Handover: by definition, a mobile moves at any time. The cellular network must be able to ensure the communication continuity when a mobile station is moving toward a new cell.

A detailed description of the problem as well as the instance used in our experiments are given in [65]. The instance consists of a 39 km wide, 168 km long highway area with 250 candidate sites (see Figure 2.8).

Performance Metrics

Two complementary performance indicators have been used for evaluating the quality of the obtained nondominated sets: the entropy and the contribution. The entropy indicator gives an idea about the diversity of the solutions found, whereas the contribution indicator compares two fronts in terms of dominance. Let PO_1 and PO_2 be the Pareto fronts of a given MOP when it is solved with two algorithms A and B, respectively. Let be $PO = ND\,(PO_1 \bigcup PO_2)$, with ND representing the nondominated set. Then, the N-dimensional space, where N is the number of objective functions to optimize, is clustered. For each space unit with at least one element of PO, the number of present solutions of PO_1 is calculated. This way, the relative entropy, $E(PO_1, PO_2)$, of a set of nondominated solutions PO_1 regarding to the Pareto frontier PO is defined as:

$$E(PO_1, PO_2) = \frac{-1}{C \log(C)} \sum_{i=1}^{C} \left(\frac{n_i}{C} \log \frac{n_i}{C} \right) \qquad (2.1)$$

where C is the cardinality of the non-empty space units of PO, and n_i is the number of solutions of set PO_1 inside the corresponding space unit. The more diversified the solution set PO_1 on the frontier the higher the entropy value $(0 \leq E \leq 1)$.

The contribution metric quantifies the dominance between two Pareto optimal sets. The contribution of PO_1 relative to PO_2 is roughly the ratio of

Fig. 2.8. CRND problem instance used

nondominated solutions produced by PO_1. This way, let C be the set of solutions in $PO_1 \cap PO_2$. Let W_1 (resp. W_2) be the set of solutions in PO_1 that dominate solutions of PO_2 (resp. in PO_1). Similarly, let L_1 (resp. L_2) be the set of solutions in PO_1 (resp. PO_2) that are dominated by solutions of PO_2 (resp. PO_1). The set of solutions in PO_1 (resp. PO_2) that are comparable to solutions in PO_2 (resp. PO_1) is $N_1 = PO_1 \backslash (C \cup W_1 \cup L_1)$ (resp. $N_2 = PO_2 \backslash (C \cup W_2 \cup L_2)$). This way, the contribution, $CONT(PO_1/PO_2)$ is stated as:

$$CONT(PO_1/PO_2) = \frac{\frac{|C|}{2} + |W_1| + |N_1|}{|C| + |W_1| + |N_1| + |W_2| + |N_2|} \qquad (2.2)$$

These two metrics allow measurements to be performed between exactly two Pareto fronts. However, when one deals with heuristic algorithms, K independent runs should be executed in order to provide the results with statistical confidence (K is typically greater than 30). In this case, for each multiobjective heuristic algorithm K Pareto fronts are obtained, so we have extended the definitions of entropy and contribution to address this fact. Let PO^A be a Pareto front obtained with the heuristic algorithm A and let $\{PO_i^B\}_{i \in \{1,...,K\}}$ be the K fronts resulting from the K independent runs of algorithms B. The extended entropy, E^{ext}, and the extended contribution, $CONT^{ext}$, are defined as:

$$E^{ext}\left(PO^A, \{PO_i^B\}_{i \in \{1,...,K\}}\right) = \frac{1}{K} \sum_{j=1}^{K} E(PO^A, PO_j^B) \quad (2.3)$$

$$CONT^{ext}\left(PO^A/\{PO_i^B\}_{i \in \{1,...,K\}}\right) = \frac{1}{K} \sum_{j=1}^{K} CONT(PO^A/PO_j^B) \quad (2.4)$$

That is, these extended metrics are the average values of E and $CONT$ of a given nondominated set over the whole set of Pareto fronts passed as the second argument.

Results

We have tested three configuration of pPAES: pPAES$_4$, pPAES$_8$, and pPAES$_{16}$, that use 4, 8, and 16 processes, respectively. Each process runs in a dedicate processor, so 4, 8, and 16 machines have been used for the pPAES experiments. The three versions of pPAES, as well as the PAES algorithm, perform 10,000 function evaluations. This way, each parallel process of pPAES$_4$ evaluates 2500 individuals, 1250 function evaluations are computed by pPAES$_8$ processes, and a process of pPAES$_{16}$ does 625 evaluations. The maximum archive size for PAES and for a pPAES island are both 100 solutions.

The encoding scheme adopted and the mutation operator applied vary depending on the problem being solved. For the mathematical test functions, all the algorithms have used binary-coded variables and the mutation is the Bit-Flip operator. The mutation rate applied is $\frac{1}{L}$, where L is the length of the binary string. As to the CRND problem, encoding and mutation operator are the same as those used in [65], whereas the mutation rate has been set to 0.05. All the values shown in the tables of this section are the average values over 30 independent runs.

Let us begin by analyzing the execution times of the algorithms. The two mathematical test MOPs, Fonseca and Kursawe, have not been considered here because their execution times are lower than one second in a single processor machine of our cluster. This way, their parallel execution times will not give us relevant information, and we just include them to evaluate the new model of search proposed by pPAES.

As to the CRDN, with the PAES configuration above, its average execution time is 19,777 seconds, i.e., 5.49 hours of computation for only 10,000 function evaluations. This fact indicates how costly the function evaluation is for this MOP. When 4, 8, and 16 processors are used, then pPAES$_4$, pPAES$_8$, and pPAES$_{16}$ last 5,153 seconds, 3,277 seconds, and 2160 seconds, respectively. Thus showing that having a higher number of processors, pPAES allows the network designer (i.e., the decision maker) to get results faster.

Table 2.4 presents the resulting values for the extended entropy and extended contribution metrics of the Pareto fronts obtained by pPAES. The E^{ext} column of this table shows the extended entropy indicator (Equation 2.3). Since E^{ext} is a relative measure between nondominated sets, the first two columns in the table contain the arguments (in order) used for calculating this metric (this is also applicable to the contribution indicator). This way, for the problems Fonseca and Kursawe, the relative entropies between PAES and the three pPAES versions (rows 1 to 3) are lower than the relative entropies between the pPAES versions and PAES (rows 4 to 6), what indicates that the new search model is able to find a more diverse Pareto front than PAES for these two mathematical MOPs. This diversity may be introduced by the migration operator which exchanges nondominated solutions among the pPAES processes. However, if we focus on the CRND problem, the algorithms yield a similar spreadout of the nondominated solutions on the frontiers (the relative entropy is around 0.42).

Table 2.4. Entropy and contribution metrics of the three considered MOPs

Arguments		E^{ext}			$CONT^{ext}$		
PO^A	$\{PO_i^B\}$	Fonseca	Kursawe	CRND	Fonseca	Kursawe	CRND
PAES	pPAES$_4$	0.3620	0.3731	0.4359	0.4251	0.4325	0.5537
PAES	pPAES$_8$	0.3617	0.3706	0.4251	0.4137	0.4237	0.5570
PAES	pPAES$_{16}$	0.3591	0.3694	0.4550	0.3674	0.4214	0.6168
pPAES$_4$	PAES	0.3957	0.4091	0.4280	0.5749	0.5675	0.4463
pPAES$_8$	PAES	0.4050	0.4116	0.4160	0.5864	0.5763	0.4431
pPAES$_{16}$	PAES	0.4206	0.4115	0.4143	0.6326	0.5786	0.3832

The contribution indicator (see the $CONT^{ext}$ column of Table 2.4) behaves slightly different. For the problems Fonseca and Kursawe, contribution values show that the three pPAES versions bring more Pareto solutions than PAES, i.e., rows 4 to 6 have larger values for this metric than rows 1 to 3 (especially pPAES$_{16}$ for the problem Fonseca, which contribution value to PAES is 0.6326). As a matter of fact, the contribution of pPAES to PAES increases with the number of processes for these two MOPs: $CONT^{ext}(pPAES_{16}/PAES) \geq CONT^{ext}(pPAES_8/PAES) \geq CONT^{ext}(pPAES_4/PAES)$. This allows us to conclude that starting the search from different points of the search space (each pPAES process begins with a random initial solution) finds more accurate solutions for these mathematical problems. Despite of the improvements yielded for the problems Fonseca and Kursawe, PAES overcomes any pPAES configuration for the CRND problem. This fact can be explained because of the algorithm configuration tested, in which the number of function evaluations performed at each pPAES island is not enough for the subalgorithms to converge towards the Pareto front of the complex real-world CRND problem. What is clear is that, once each island converges to the Pareto optimal front, pPAES provides the resulting front with a higher solution distribution (diversity).

2.4 Summary

In this chapter we have performed a wide review of the literature concerning parallel multiobjective evolutionary algorithms. A classification of the MOEAs attending to their level of parallelism and their management of the Pareto front has been presented. We have then analyzed the contributions on parallel MOEAs from an historical point of view, the main parallel programming environments used in their development, and the communication topology of these parallel algorithms. The application domain of MOPs being solved has also been discussed. We complete the review with others parallel non-MOEA approaches for solving MOPs.

Next, a parallel MOEA, named pPAES, has been implemented. This algorithm is based on PAES. We have tested pPAES with two mathematical MOPs, Fonseca and Kursawe, and a real problem coming from the telecommunication industry, the CRND problem. The two former functions have been used for showing the working principles of the algorithm, while the latter real-world MOP has been used for illustrating the actual power of our algorithm. Then, as to the CRND problem, we have shown that pPAES is able to provide the network designer (i.e., the decision maker) with (hopefully) optimal trade-off network configurations faster than PAES. From a quality of the results point of view, the larger the number of processes used in the pPAES configuration, the higher the diversity and the lower the closeness of the resulting Pareto fronts for the two mathematical MOPs.

As future research lines, we plan to deeply study pPAES on the CRND problem by increasing the number of iterations of the algorithm. This way, we will hopefully be able to improve the performance of pPAES for solving this complex MOP. We also intend to develop new parallel models of search possibly based on other multiobjective evolutionary algorithms.

Acknowledgments

This work has been partially funded by the Ministry of Science and Technology and FEDER under contracts TIC2002-04498-C05-02 (the TRACER project) and TIC2002-04309-C02-02.

References

1. D.K. Agrafiotis. Multiobjective Optimization of Combinatorial Libraries. *IBM J. RES. & DEV.*, 45(3/4):545–566, 2001.
2. A. Al-Yamani, S. Sait, and H. Youssef. Parallelizing Tabu Search on a Cluster of Heterogeneous Workstations. *Journal of Heuristics*, 8(3):277–304, 2002.
3. A. Al-Yamani, S.M. Sait, H. Barada, and H. Youssef. Parallel Tabu Search in a Heterogeneous Environment. In *Proc. of the Int. Parallel and Distributed Processing Symp.*, pages 56–63, 2003.
4. A. Al-Yamani, S.M. Sait, and H.R. Barada. HPTS: Heterogeneous Parallel Tabu Search for VLSI Placement. In *Proc. of the 2002 Congress on Evolutionary Computation*, pages 351–355, 2002.
5. E. Alba, editor. *Parallel Metaheuristics: A New Class of Algorithms*. John Wiley & Sons, 2005.
6. E. Alba and M. Tomassini. Parallelism and Evolutionary Algorithms. *IEEE Transactions on Evolutionary Computation*, 6(5):443–462, 2002.
7. E. Alba and J.M. Troya. A Survey of Parallel Distributed Genetic Algorithms. *Complexity*, 4(4):31–52, 1999.
8. G. Aloisio, E. Blasi, M. Cafaro, I. Epicoco, S. Fiore, and S. Mocavero. A Grid Environment for Diesel Engine Chamber Optimization. In *Proc. of ParCo2003*, 2003.
9. M. Basseur, J. Lemesre, C. Dhaenens, and E.-G. Talbi. Cooperation Between Branch and Bound and Evolutionary Approaches to Solve a Biobjective Flow Shop Problem. In *Workshop on Evolutionary Algorithms (WEA'04)*, pages 72–86, 2004.
10. C.P. Bottura and J.V. da Fonseca Neto. Rule-Based Decision-Making Unit for Eigenstruture Assignment via Parallel Genetic Algorithm and LQR Designs. In *Proc. of the American Control Conference*, pages 467–471, 2000.
11. J. Branke, H. Schmeck, K. Deb, and M. Reddy S. Parallelizing Multi-Objective Evolutionary Algorithms: Cone Separation. In *Congress on Evolutionary Computation (CEC'2004)*, pages 1952–2957, 2004.
12. E. Cantú-Paz. *Efficient and Accurate Parallel Genetic Algorithms*. Kluwer Academic Publisher, 2000.

13. C.S. Chang and J.S. Huang. Optimal Multiobjective SVC Planning for Voltage Stability Enhancement. *IEE Proc.-Generation, Transmission, and Distribution*, 145(2):203–209, 1998.

14. C.A. Coello and M. Reyes. A Study of the Parallelization of a Coevolutionary Multi-Objective Evolutionary Algorithm. In *MICAI 2004*, LNAI 2972, pages 688–697, 2004.

15. C.A. Coello and G. Toscano. Multiobjective Optimization Using a Micro-Genetic Algorithm. In *GECCO-2001*, pages 274–282, 2001.

16. C.A. Coello, D.A. Van Veldhuizen, and G.B. Lamont. *Evolutionary Algorithms for Solving Multi-Objective Problems*. Kluwer Academic Publishers, 2002.

17. M. Conti, S. Orcioni, and C. Turchetti. Parametric Yield Optimisation of MOS VLSI Circuits Based on Simulate Annealing and its Parallel Implementation. *IEE Proc.-Circuits Devices Syst.*, 141(5):387–398, 1994.

18. J.V. da Fonseca Neto and C.P. Bottura. Parallel Genetic Algorithm Fitness Function Team for Eigenstructure Assignment via LQR Designs. In *Proc. of the 1999 Congress on Evolutionary Computation*, pages 1035–1042, 1999.

19. A. de Risi, T. Donateo, D. Laforgia, G. Aloisio, E. Blasi, and S. Mocavero. An Evolutionary Methodology for the Design of a D.I. Combustion Chamber for Diesel Engines. In *Conf. on Thermo- and Fluid Dynamics Processes in Diesel Engines (THIESEL 2004)*, 2004.

20. F. de Toro, J. Ortega, J. Fernández, and A. Díaz. PSFGA: A Parallel Genetic Algorithm for Multiobjective Optimization. In *Proc. of the 10th Euromicro Workshop on Parallel, Distributed and Network-Based Processing*, pages 384–391, 2002.

21. F. de Toro, J. Ortega, and B. Paechter. Parallel Single Front Genetic Algorithm: Performance Analysis in a Cluster System. In *Proc. of the Int. Parallel and Distributed Processing Symp. (IPDPS'03)*, page 143, 2003.

22. F. de Toro, J. Ortega, E. Ros, S. Mota, B. Paechter, and J.M. Martín. PSFGA: Parallel Processing and Evolutionary Computation for Multiobjective Optimisation. *Parallel Computing*, 30(5-6):721–739, 2004.

23. K. Deb. *Multi-Objective Optimization Using Evolutionary Algorithms*. Wiley, 2001.

24. K. Deb, A. Pratap, S. Agarwal, and T. Meyarivan. A Fast and Elitist Multiobjective Genetic Algorithm: NSGA-II. *IEEE Trans. on Evolutionary Computation*, 6(2):182–197, 2002.

25. K. Deb, P. Zope, and A. Jain. Distributed Computing of Pareto-Optimal Solutions Using Multi-Objective Evolutionary Algorithms. KanGAL 2002008, Indian Institute of Technology Kampur, 2002.

26. K. Deb, P. Zope, and A. Jain. Distributed Computing of Pareto-Optimal Solutions Using Multi-Objective Evolutionary Algorithms. In *EMO 2003*, LNCS 2632, pages 534–549, 2003.

27. P. Delisle, M. Krajecki, M. Gravel, and C. Gagn. Parallel Implementation of an Ant Colony Optimization Metaheuristic with OpenMP. In *Proc. of the 3rd European Workshop on OpenMP (EWOMP01)*, 2001.

28. D.J. Doorly and J. Peiró. Supervised Parallel Genetic Algorithms in Aerodynamic Optimisation. In *Proc. of the 13th AIAA CFD Conference*, pages 210–216, 1997.

29. D.J. Doorly, J. Peiró, and J.-P. Oesterle. Optimisation of Aerodynamic and Coupled Aerodynamic-Structural Design Using Parallel Genetic Algorithms. In

   Proc. of the Sixth AIAA/NASA/USAF Multidiscipliary Analysis and Optimization Symp., pages 401–409, 1996.
30. D.J. Doorly, S. Spooner, and J. Peiró. Supervised Parallel Genetic Algorithms in Aerodynamic Optimisation. In *EvoWorkshops 2000*, LNCS 1803, pages 357–366, 2000.
31. S. Duarte and B. Barán. Multiobjective Network Design Optimisation Using Parallel Evolutionary Algorithms. In *XXVII Conferencia Latinoamericana de Informática CLEI'2001*, 2001.
32. P. Eberhard, F. Dignath, and L. Kübler. Parallel Evolutionary Optimization of Multibody Systems with Application to Railway Dynamics. *Multibody Systems Dynamics*, 9:143–164, 2003.
33. C.M. Fonseca and P. Fleming. Genetic Algorithms for Multiobjective Optimization: Formulation, Discussion and Generalization. In *Proc. of the 5th Int. Conf. on Genetic Algorithms*, pages 416–423, San Mateo, CA, 1993.
34. C.M. Fonseca and P.J. Flemming. Multiobjective Optimization and Multiple Constraint Handling with Evolutionary Algorithms – Part II: Application Example. *IEEE Transactions on System, Man, and Cybernetics*, 28:38–47, 1998.
35. A. Geist, W. Jiang R. Manchek A. Beguelin, J. Dongarra, and V. Sunderam. *PVM: Parallel Virtual Machine*. The MIT Press, 1994.
36. T. Goel, R. Vaidyanathan, R. Haftka, and W. Shyy. Response Surface Approximation of Pareto Optimal Front in Multi-objective Optimization. In *10th AIAA/ISSMO Multidisciplinary Analysis and Optimization Conference*, 2004.
37. I.E. Golovkin, S.J. Louis, and R.C. Mancini. Parallel Implementation of Niched Pareto Genetic Algorithm Code for X-Ray Plasma Spectroscopy. In *Proc. of the 2002 Congress on Evolutionary Computation*, pages 1820–1824, 2002.
38. I.E. Golovkin, R.C. Mancini, and S.J. Louis. Parallel Implementation of Niched Pareto Genetic Algorithm Code for X-Ray Plasma Spectroscopy. In *Late-Breaking Papers at the 2000 Genetic and Evolutionary Computation Conference*, 2000.
39. W. Gropp, E. Lusk, and A. Skjellum. *Using MPI: Portable Parallel Programming with the Message-passing Interface*. The MIT Press, London, UK, 2000.
40. T. Hiroyasu. Diesel Engine Design Using Multi-Objective Genetic Algorithm. In *Japan/US Workshop on Design Environment 2004*, 2004.
41. T. Hiroyasu, M. Miki, M. Kim, S. Watanabe, H. Hiroyasu, and H. Miao. Reduction of Heavy Duty Diesel Engine Emission and Fuel Economy with Multi-Objective Genetic Algorithm and Phenomenological Model. *SAE Paper SP-1824*, 2004.
42. T. Hiroyasu, M. Miki, and S. Watanabe. Divided Range Genetic Algorithms in Multiobjective Optimization Problems. In *Proc. of Int. Workshop on Emergent Synthesis, IWES' 99*, pages 57–66, 1999.
43. T. Hiroyasu, M. Miki, and S. Watanabe. The New Model of Parallel Genetic Algorithm in Multi-Objective Optimization Problems – Divided Range Multi-Objective Genetic Algorithm –. In *2000 IEEE Congress on Evolutionary Computation*, pages 333–340, 2000.
44. H. Horii, M. Miki, T. Koizumi, and N. Tsujiuchi. Asynchronous Migration of Island Parallel GA For Multi-Objective Optimization Problem. In *Proc. of the 4th Asia-Pacific Conf. on Simulated Evolution and Learning (SEAL'02)*, pages 86–90, 2002.

45. J. Horn, N. Nafpliotis, and D.E. Goldberg. A Niched Pareto Genetic Algorithm for Multi-Objective Optimization. In *Proc. of the 1st IEEE Conf. on Evolutionary Computation*, pages 82–87, 1994.
46. D. Jaeggi, C. Asselin-Miller, G. Parks, T. Kipouros, T. Bell, and J. Clarkson. Multi-Objective Parallel Tabu Search. In *Parallel Problem Solving from Nature (PPSN VIII)*, LNCS 3242, pages 732–741, 2004.
47. D. Jaeggi, G. Parks, T. Kipouros, and J. Clarkson. A Multi-Objective Tabu Search Algorithm for Constrained Optimisation Problems. In *Third Int. Conf. on Evolutionary Multi-Criterion Optimization (EMO'05)*, LNCS 3410, pages 490–504, 2005.
48. B.R. Jones, W.A. Crossley, and A.S. Lyrintzis. Aerodynamic and Aeroacoustic Optimization of Airfoils Via a Parallel Genetic Algorithm. In *Proc. of the 7th AIAA/USAF/NASA/ISSMO Symp. on Multidisciplinary Analysis and Optimization*, 1998.
49. N. Jozefowiez, F. Semet, and E.-G. Talbi. Parallel and Hybrid Models for Multi-Objective Optimization: Application to the Vehicle Routing Problem. In *Parallel Problem Solving from Nature (PPSN VII)*, pages 271–280, 2002.
50. J. Kamiura, T. Hiroyasu, M. Miki, and S. Watanabe. MOGADES: Multi-Objective Genetic Algorithm with Distributed Environment Scheme. In *Proc. of the 2nd Int. Workshop on Intelligent Systems Design and Applications (ISDA'02)*, pages 143–148, 2002.
51. M. Kanazaki, M. Morikawa, S. Obayashi, and K. Nakahashi. Multiobjective Design Optimization of Merging Configuration for an Exhaust Manifold of a Car Engine. In *Parallel Problem Solving from Nature (PPSN VII)*, pages 281–287, 2002.
52. N. Keerativuttitumrong, C. Chaiyaratana, and V. Varavithya. Multi-Objective Co-Operative Co-Evolutionary Genetic Algorithm. In *Parallel Problem Solving from Nature (PPSN VII)*, pages 288–297, 2002.
53. M.P. Kleeman, R.O. Day, and G.B. Lamont. Analysis of a Parallel MOEA Solving the Multi-Objective Quadratic Assignment Problem. In *GECCO 2004*, LNCS 3103, pages 402–403, 2004.
54. M.R. Knarr, M.N. Goltz, G.B. Lamont, and J. Huang. *In Situ* Bioremediation of Perchlorate-Contaminated Groundwater Using a Multi-Objective Parallel Evolutionary Algorithm. In *Proc. of the 2003 Congress on Evolutionary Computation (CEC'2003)*, pages 1604–1611, 2003.
55. J.D. Knowles and D.W. Corne. Approximating the Nondominated Front Using the Pareto Archived Evolution Strategy. *Evolutionary Computation*, 8(2):149–172, 2000.
56. F. Kursawe. A Variant of Evolution Strategies for Vector Optimization. In H.P. Schwefel and R. Männer, editors, *Parallel Problem Solving for Nature*, pages 193–197, Berlin, Germany, 1990. Springer-Verlag.
57. J. Lemesre, C. Dhaenens, and E.-G. Talbi. A Parallel Exact Method for a Bicriteria Permutation Flow-Shop Problem. In *Project Management and Scheduling (PMS'04)*, pages 359–362, 2004.
58. J. Lienig. A Parallel Genetic Algorithm for Performance-Driven VLSI Routing. *IEEE Trans. on Evolutionary Computation*, 1(1):29–39, 1997.
59. D.A. Linkens and H. Okola Nyongesa. A Distributed Genetic Algorithm for Multivariable Fuzzy Control. In *IEE Colloquium on Genetic Algorithms for Control Systems Engineering*, pages 9/1–9/3, 1993.

60. R.A.E. Mäkinen, P. Neittaanmäki, J. Periaux, M. Sefrioui, and J. Toivanen. Parallel Genetic Solution for Multobjective MDO. In *Parallel CFD'96 Conference*, pages 352–359, 1996.
61. J.M. Malard, A. Heredia-Langner, D.J. Baxter, K.H. Jarman, and W.R. Cannon. Constrained De Novo Peptide Identification via Multi-Objective Optimization. In *Online Proc. of the Third IEEE Int. Workshop on High Performance Computational Biology (HiCOMB 2004)*, 2004.
62. S. Manos and L. Poladian. Novel Fibre Bragg Grating Design Using Multiobjective Evolutionary Algorithms. In *Proc. of the 2003 Congress on Evolutionary Computation (CEC'2003)*, pages 2089–2095, 2003.
63. N. Marco and S. Lanteri. A Two-Level Parallelization Strategy for Genetic Algorithms Applied to Optimum Shape Design. *Parallel Computing*, 26:377–397, 2000.
64. J. Mehnen, T. Michelitsch, K. Schmitt, and T. Kohlen. pMOHypEA: Parallel Evolutionary Multiobjective Optimization using Hypergraphs. Technical Report CI-187/05, Universität Dortmund, 2005.
65. H. Meunier, E.-G. Talbi, and P. Reininger. A Multiobjective Genetic Algorithm for Radio Network Design. In *Proc. of the 2000 Congress on Evolutionary Computation*, pages 317–324, 2000.
66. A.J. Nebro, E. Alba, and F. Luna. Multi-Objective Optimization Using Grid Computing. *Soft Computing Journal*, 2005. To appear.
67. S. Obayashi, D. Sasaki, Y. Takeguchi, and N. Hirose. Multiobjective Evolutionary Computation for Supersonic Wing-Shape Optimization. *IEEE Trans. on Evolutionary Computation*, 4(2):182–187, 2000.
68. S. Obayashi, T. Tsukahara, and T. Nakamura. Cascade Airfoil Design by Multiobjective Genetic Algorithms. In *Second Int. Conf. on Genetic Algorithms in Engineering Systems: Innovations and Applications*, pages 24–29, 1997.
69. S. Obayashi, T. Tsukahara, and T. Nakamura. Multiobjective Genetic Algorithm Applied to Aerodynamic Design of Cascade Arfoils. *IEEE Trans. on Industrial Electronics*, 47(1):211–216, 2000.
70. C.K. Oh and G.J. Barlow. Autonomous Controller Design for Unmanned Aerial Vehicles using Multi-objective Genetic Programming. In *Proceedings of the 2004 IEEE Congress on Evolutionary Computation*, pages 1538–1545, 2004.
71. L.S. Oliveira, R. Sabourin, F. Bortolozzi, and C.Y. Suen. A Methodology for Feature Selection Using Multiobjective Genetic Algorithms for Handwritten Digit String Recognition. *Int. Journal of Pattern Recognition and Artificial Intelligence*, 17(6):903–929, 2003.
72. A. Osyczka. *Multicriteria Optimization for Engineering Design*. Academic Press, 1985.
73. K.E. Parsopoulos, D.K. Tasoulis, N.G. Pavlidis, V.P. Plagianakos, and M.N. Vrahatis. Vector Evaluated Differential Evolution for Multiobjective Optimization. In *Proc. of the IEEE 2004 Congress on Evolutionary Computation (CEC 2004)*, pages 204–211, 2004.
74. K.E. Parsopoulos, D.K. Tasoulis, and M.N. Vrahatis. Multiobjective Optimization Using Parallel Vector Evaluated Particle Swarm Optimization. In *Proc. of the IASTED Int. Conf. on Artificial Intelligence and Applications (AIA 2004)*, pages 823–828, 2004.
75. C. Poloni, A. Giurgevich, L. Onesti, and V. Pediroda. Hybridization of a Multi-Objective Genetic Algorithm, a Neural Network and a Classical Optimizer for

a Complex Design Problem in Fluid Dynamics. *Computer Methods in Applied Mechanics and Engineering*, 186:403–420, 2000.

76. D. Quagliarella and A. Vicini. Sub-Population Policies for a Parallel Multiobjective Genetic Algorithm with Application to Wing Design. In *1998 IEEE Int. Conf. On Systems, Man, And Cybernetics*, pages 3142–3147, 1998.

77. P.W.W. Radtke, L.S. Oliveira, R. Sabouring, and T. Wong. Intelligent Zoning Design Using Multi-Objective Evolutionary Algorithms. In *Proc. of the Seventh Int. Conf. on Document Analysis and Recognition (ICDAR 2003)*, pages 824–828, 2003.

78. J.L. Rogers. A Parallel Approach to Optimum Actuator Selection with a Genetic Algorithm. In *AIAA Guidance, Navigation, and Control Conf.*, 2000.

79. J. Rowe, K. Vinsen, and N. Marvin. Parallel GAs for Multiobjective Functions. In *Proc. of the 2nd Nordic Workshop on Genetic Algorithms and Their Applications (2NWGA)*, pages 61–70, 1996.

80. S.M. Sait, H. Youssef, H.R. Barada, and A. Al-Yamani. A Parallel Tabu Search Algorithm for VLSI Standard-Cell Placement. In *ISCAS'00*, pages 581–584, 2000.

81. D. Sasaki, M. Morikawa, S. Obayashi, and K. Nkahashi. Aerodynamic Shape Optimization of Supersonic Wings by Adaptive Range Multiobjective Genetic Algorithms. In *First Int. Conf. on Evolutionary Multi-Criterion Optimization (EMO 2001)*, 2001.

82. J.D. Schaffer. *Multiple Objective Optimization with Vector Evaluated Genetic Algorithms.* PhD thesis, Vanderbilt University, Nashville, TN, USA, 1984.

83. T.J. Stanley and T. Mudge. A Parallel Genetic Algorithm for Multiobjetive Microprocessor Design. In *Proc. of the Sixth Int. Conf. on Genetic Algorithms*, pages 597–604, 1995.

84. G. Stehr, H. Graeb, and K. Antreich. Performance Trade-off Analysis of Analog Circuits by Normal-Boundary Intersection. In *Proc. of the 40th Conference on Design automation*, pages 958–963, 2003.

85. F. Streichert, H. Ulmer, and A. Zell. Parallelization of Multi-Objective Evolutionary Algorithms Using Clustering Algorithms. In *Third Int. Conf. on Evolutionary Multi-Criterion Optimization (EMO'05)*, LNCS 3410, pages 92–107, 2005.

86. K.C. Tan, Y.J. Yang, and T.H. Lee. A Distributed Cooperative Coevolutionary Algorithm for Multiobjective Optimization. In *Proc. of the 2003 Congress on Evolutionary Computation (CEC'2003)*, pages 2513–2520, 2003.

87. D.A. Van Veldhuizen and G.B. Lamont. Multiobjective Evolutionary Algorithm Test Suites. In *Proc. of the 1999 ACM Symp. on Applied Computing*, pages 351–357, 1999.

88. A. Vicini and D. Quagliarella. A Multiobjective Approach to Transonic Wing Design by Means of Genetic Algorithms. In *NATO RTO AVT Symposium on Aerodynamic Design and Optimization*, 1999.

89. J.F. Wang, J. Periaux, and M. Sefrioui. Parallel Evolutionary Algorithms for Optimization Problems in Aerospace Engineering. *Journal of Computational and Applied Mathematics*, 149:155–169, 2002.

90. S. Watanabe, T. Hiroyasu, and M. Miki. Parallel Evolutionary Multi-Criterion Optimization for Block Layout Problems. In *2000 Int. Conf. on Parallel and Distributed Processing Techniques and Applications (PDPTA'2000)*, pages 667–673, 2000.

56   Francisco Luna, Antonio J. Nebro, and Enrique Alba

91. S. Watanabe, T. Hiroyasu, and M. Miki. Parallel Evolutionary Multi-Criterion Optimization for Mobile Telecommunication Networks Optimization. In *Proc. of the EUROGEN'2001*, pages 167–172, 2001.
92. M.M. Wiecek and H. Zhang. A Scalable Parallel Algorithm for Multiple Objective Linear Programs. ICASE 94-38, NASA, 1994.
93. N. Xiao and M. P. Armstrong. A Specialized Island Model and Its Applications in Multiobjective Optimization. In *Genetic and Evolutionary Computation Conference (GECCO'03)*, LNCS 2724, pages 1530–1540, London, UK, 2003. Springer-Verlag.
94. S. Xiong and F. Li. Parallel Strength Pareto Multi-Objective Evolutionary Algorithm. In *Proc. of the 2003 Congress on Evolutionary Computation (CEC'2003)*, pages 681–683, 2003.
95. D. Zaharie and D. Petcu. Adaptive Pareto Differential Evolution and Its Parallelization. In *PPAM 2003*, LNCS 3019, pages 261–268, 2004.
96. Z.-Y. Zhu and K.-S. Leung. Asynchronous Self-Adjustable Island Genetic Algorithm for Multi-Objective Optimization Problems. In *Proc. of the 2002 Congress on Evolutionary Computation (CEC'2002)*, pages 837–842, 2002.
97. E. Zitzler, M. Laumanns, and L. Thiele. SPEA2: Improving the Strength Pareto Evolutionary Algorithm. Technical Report 103, Computer Engineering and Networks Laboratory (TIK), Swiss Federal Institute of Technology (ETH), Zurich, Switzerland, 2001.

Part II

Parallel Hardware for Genetic Algorithms

3

A Reconfigurable Parallel Hardware for Genetic Algorithms

Nadia Nedjah and Luiza de Macedo Mourelle

[1] Department of Electronics Engineering and Telecommunications,
Faculty of Engineering, State University of Rio de Janeiro, Brazil
nadia@eng.uerj.br, http://www.eng.uerj.br/~nadia/english.html
[2] Department of Systems Engineering and Computation,
Faculty of Engineering, State University of Rio de Janeiro, Brazil
ldmm@eng.uerj.br, http://www.eng.uerj.br/~ldmm

In this chapter, we propose a massively parallel architecture of a hardware implementation of genetic algorithms. This design is quite innovative as it provides a viable solution to the fitness computation problem, which depends heavily on the problem-specific knowledge. The proposed architecture is completely independent of such specifics. It implements the fitness computation using a neural network. The hardware implementation of the used neural network is stochastic and thus minimises the required hardware area without much increase in response time. Last but not least, we demonstrate the characteristics of the proposed hardware and compare it to existing ones.

3.1 Introduction

Generally speaking, a *genetic algorithm* is a process that evolves a set of *individuals*, also called *chromosomes*, which constitutes the *generational population*, producing a new population. The individuals represent a solution to the problem in consideration. The freshly produced population is yield using some genetic operators such as *selection*, *crossover* and *mutation* that attempt to simulate the natural breeding process in the hope of generating new solutions that are *fitter*, i.e. adhere more the problem constraints.

Previous work on hardware genetic algorithms can be found in [5, 10, 12]. Mainly, Earlier designs are hardware/software codesigns and they can be divided into three distinct categories: *(i)* those that implement the fitness computation in hardware and all the remaining steps including the genetic operators in software, claiming that the bulk computation within genetic evolution is the fitness computation. The hardware is problem-dependent; *(ii)* and those

N. Nedjah and L. de Macedo Mourelle: *A Reconfigurable Parallel Hardware for Genetic Algorithms*, Studies in Computational Intelligence (SCI) **22**, 59–69 (2006)
www.springerlink.com   © Springer-Verlag Berlin Heidelberg 2006

that implement the fitness computation in software and the rest in hardware, claiming that the ideal candidate are the genetic operators as these exhibit regularity and generality [2, 7]. *(iii)* those that implement the whole genetic algorithm in hardware [10]. We believe that both approaches are worthwhile but a hardware-only implementation of both the fitness calculation and genetic operators is also valuable. Furthermore, a hardware implementation that is problem-independent is yet more useful.

The remainder of this chapter is divided into five sections. In Section 3.2, we describe the principles of genetic algorithms. Subsequently, in Section 3.3, we propose and describe the overall hardware architecture of the problem-independent genetic algorithm. Thereafter, in Section 3.4, we detail the architecture of each of the component included in the hardware genetic algorithm proposed. Then, in Section 3.5, assess the performance of the proposed architecture. Finally, we draw some conclusions 3.6.

3.2 Principles of Genetic Algorithms

Genetic algorithms maintain a *population* of *individuals* that evolve according to *selection rules* and other *genetic operators*, such as *mutation* and *crossover*. Each individual receives a measure of *fitness*. Selection focuses on high fitness individuals. Mutation and crossover provide general heuristics that simulate the reproduction process. Those operators attempt to perturb the characteristics of the parent individuals as to generate distinct offspring individuals.

Genetic algorithms are implemented through the procedure described by Algorithm 1, wherein parameters *ps*, *ef* and *gn* are the population size, the expected fitness of the returned solution and the maximum number of generation allowed respectively.

Algorithm 1. GA – Genetic algorithms basic cycle
Input: population size *ps*, expected fitness *ef*, generation number *gn*;
Output: the problem solution;
1:   generation := 0;
2:   population := initialPopulation();
3:   fitness := evaluate(population);
4:   **do**
5:   parents := select(population);
6:   population := mutate(crossover(parents));
7:   fitness := evaluate(population);
8:   generation := generation + 1;
9:   **while** (fitness[i] $\neq$ *ef*, $1 \leq i \leq ps$) and (generation < *gn*);
10:   **return** fittestIndividual(population);
End.

In Algorithm 1, function *intialPopulation* returns a valid random set of individuals that compose the population of first generation while function *evaluate* returns the fitness of a given population storing the result into fitness. Function *select* chooses according to some random criterion that privilege fitter individuals, the individuals that should be used to generate the population of the next generation and function *crossover* and *mutate* implement the crossover and mutation process respectively to actually yield the new population.

3.3 Overall Architecture for the Hardware Genetic Algorithm

Clearly, for hardware genetic algorithms, individuals are always represented using their binary representation. Almost all aspects of genetic algorithms are very attractive for hardware implementation. The selection, crossover and mutation processes are generic and so are problem-independent. The main issue in the hardware implementation of genetic algorithms is the computation of individual's fitness values. This computation depends on problem-specific knowledge. The novel contribution of the work consists of using neural network hardware to compute the fitness of individuals. The software version of the neural network is trained with a variety of individual examples. Using a hardware neural network to compute individual fitness yields a hardware genetic algorithm that is fully problem-independent.

The overall architecture of the proposed hardware is given Fig. 3.1. It is massively parallel. The selection process is performed in one clock cycle while the crossover and mutation processes are completed within two clock cycles.

The fitness of individual in the generational population is evaluated using hardware neural networks, which take advantage of stochastic representation of signals to reduce the hardware area required [9]. Stochastic computing principles are well detailed in [4]. The motivation behind the use of stochastic arithmetic is its simplicity. Designers are faced with hardware implementations that are very large due to large digital multipliers, adders, etc.. Stochastic arithmetic provides a way of performing complex computations with very simple hardware. Stochastic arithmetic provides a very low computation hardware area and fault tolerance. Adders and subtracters can be implemented by an ensemble of multiplexers and multipliers by a series of XOR gates. (For formal proofs on stochastic arithmetic, see [3, 9].)

3.4 Detailed Component Architectures

In this section, we concentrate on the hardware architecture of the components included in the overall architecture of Fig. 3.1.

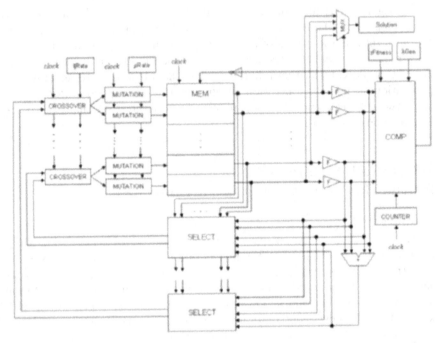

Fig. 3.1. Overall architecture of the hardware genetic algorithm proposed

3.4.1 Shared Memory for Generational Population

The generational population is kept in a synchronised bank of registers that can be read and updated. The basic element of the shared memory is an individual. An individual is a simply a bit stream of fixed size. The memory is initially filled up with a fixed initial population. Each time the comparator's output (i.e. component COMP) is 0, the content of the registers changes with the individuals provided as inputs. This happens whenever the fitness f of the best individual of the current generation is not as expected (i.e., $f > \epsilon Fitness$) and the current generation g is not the last allowed one (i.e., $g \neq \lambda Gen$). Note that $\epsilon Fitness$ and λGen are two registers that store the expected fitness value and the maximum number of generation allowed respectively.

3.4.2 Random Number Generator

A central component to the proposed hardware architecture of genetic algorithms is a source of pseudorandom noise. A source of pseudorandom digital noise consists of a *linear feedback shift register* or LFSR, described by first in [1] and by many others, for instance [3], LFSRs are very practical as they can easily be constructed using standard digital components.

Linear feedback shift registers can be implemented in two ways. The Fibonacci implementation consists of a simple shift register in which a binary-weighted modulo-2 sum of the taps is fed back to the input. Recall that modulo-2 sum of two one-bit binary numbers yields 0 if the two numbers are identical and 1 if not. The Galois implementation consists of a shift register, the content of which is modified at every step by a binary-weighted value of the output stage. The architecture of the LFSR using these methods are shown in Fig. 3.2.

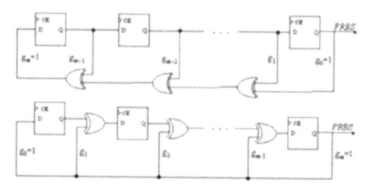

Fig. 3.2. Pseudorandom bitstream generators – Fibonacci vs. Galois implementation

Left feedback shift registers such as those of Fig. 3.2 can be used to generate multiple pseudorandom bit sequences. However, the taps from which these sequences are yield as well as the length of the LFSR must be carefully chosen. (See [3, 4] for possible length/tap position choices).

3.4.3 Selection Component

The selection component implements a variation of the roulette wheel selection. The interface of this component consists of all the individuals, say i_1, i_2, ..., i_{n-1}, i_n of the generational population of size n together with the respective fitness, say f_1, f_2, ..., f_{n-1}, f_n and the overall sum of all these fitness values, say sum. The component proceeds as described in the following steps:

1. A random number, say ρ is generated;
2. The sum of the individual's fitness values is scaled down using ρ, i.e
   $ssum := sum - \rho$;
3. Choose an individual, say i_j from the selection pool and cumulate the corresponding fitness f_j, i.e. $csum := csum + f_j$;
4. Compare the scaled sum and the so far cumulated sum and select individual i_j if $csum > ssum$, otherwise go back to step 1;

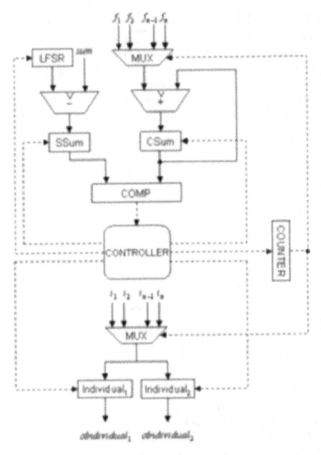

Fig. 3.3. The architecture of the selection component

5. When the first individual is selected, go back to step 1 and apply the same process to select the second individual.

The architecture of the selection component is shown in Fig. 3.3. The above iterative process is implemented using a state machine (CONTROLLER in Fig. 3.3). The state machine has 6 states and the associated state transition function is described in Fig. 3.4. The actions performed in each state of the controller machine are described below. Signal *compare* is set when the either an individual having the expected fitness is found or the last generation has passed.

S_0:   initialise counter;
         load register CSUM with 0;
S_1:   stop the random number generator;
S_2:   load register SSUM;

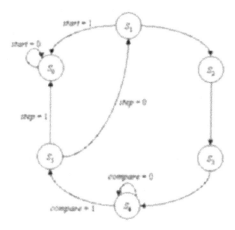

Fig. 3.4. The state transition function of the selection component controller

S_3:   load register CSUM;
S_4:   if *compare* $= 1$ then
                if *step* $= 0$ then
                        load register INDIVIDUAL$_1$;
                        start the random number generator;
                else load register INDIVIDUAL$_2$;
                increment the counter;
S_5:   if *step* $= 0$ then set *step*;

3.4.4 Genetic Operator's Components

The genetic operators are the crossover followed by the mutation. The crossover component implements the double-point crossover. It uses a linear feedback shift register which provides the random number that allows the component to decide whether to actually perform the crossover or not. This depends on whether the randomised number surpasses the informed crossover rate $\xi Rate$. In the case it does, the bits of the less significant half of the randomised number is used as the first crossover point and the most significant part as the second one.

The mutation component also uses a random number generator. The generated number must be bigger that the given mutation rate $\mu Rate$ for the mutation to occur. The bits of the randomised number are also used as way to choose the mutation degree of the individual. Starting from the less significant bit of both the random number and the individual, if the bit in the former is 1 then the corresponding bit in the later is complemented and otherwise it is kept unchanged. The hardware architecture of the mutation component is given in Fig. 3.6.

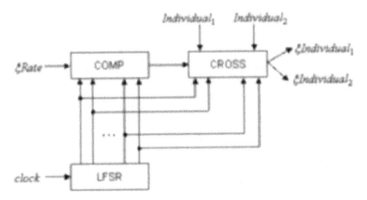

Fig. 3.5. The architecture of the crossover component

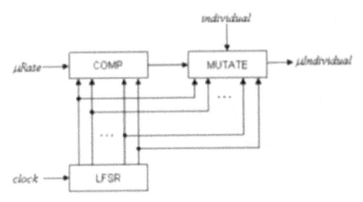

Fig. 3.6. The architecture of the crossover component

3.4.5 Fitness Evaluation Component

The individual fitness measure is estimated using neural networks. In previous work, the authors proposed and implemented a hardware for neural networks [9]. The implementation uses stochastic signals and therefore reduces very significantly the hardware area required for the network. The network topology used is the fully-connected feed-forward. The neuron architecture is given in Fig. 3.7. (More details can be found in [9].) For the hardware genetic implementation, the number of input neurons is the same as the size of the individual. The output neuron are augmented with a shift register to store the final result. The training phase is supposed to be performed before the first use within the hardware genetic algorithm.

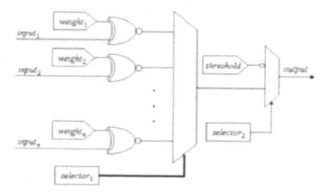

Fig. 3.7. Stochastic bipolar neuron architecture ([9])

3.5 Performance Results

The hardware genetic algorithm proposed was simulated then programmed
into an Spartan3 Xilinx FPGA [13]. In order to assess the performance of
the proposed hardware genetic algorithm, we maximise the function that was
first used in [8]. It was also used by Scott, Seth and Samal to evaluate their
hardware implementation for genetic algorithms [11]. The function is not easy
to maximise, which is clear from the function plot of Fig. 3.8. The training
phase of the neural network was done by software using toolbox offered in
MatLab [6].

$$f(x, y) = 21.5 + x\sin(4\pi x) + y\sin(20\pi y),$$

$$-3.0 \leq x_1 \leq 12.1$$
$$4.1 \leq x_2 \leq 5.8$$

(3.1)

The characteristics of the software and hardware implementations proposed in [11] and those of the hardware genetic algorithm we proposed in this
chapter are compared in Table 3.1. It is clear that the hardware implementations are both much faster than the software version. One can clearly note that
our implementation (PHGA) requires more than twice that required by Scott,
Seth and Samal's implementation (HEGA). Note, however, that the hardware
area necessary to the computation of the fitness function is not included as it
is not given in [11]. From another perspective, PHGA is more than five time
faster as it is massively parallel. We also believe that the computation of the
fitness function is much faster with the neural network. Observe that PHGA
evolved a better solution.

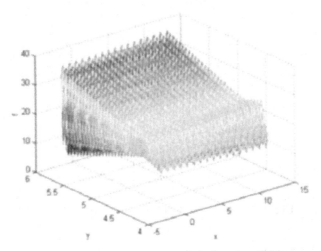

Fig. 3.8. Plotting Michalewics's function ([8])

Table 3.1. Comparison of the performance results: software genetic algorithms (SGA), hardware engine for genetic algorithms (HEGA) and proposed hardware genetic algorithms (PHGA). (The area is expressed in terms of CLBs and the time is in seconds.)

Implementation	time	area	solution	x	y	area × time
SGA	40600	0	38.5764	–	–	–
HEGA	972	870	38.8419	–	–	845640
PHGA	189	1884	38.8483	11.6241	5.7252	356076

3.6 Summary

In this chapter, we proposed a novel hardware architecture for genetic algorithms. It is novel in the sense that is massively parallel and problem-independent. It uses neural networks to compute the fitness measure. Of course, for each type of problem, the neuron weights need to be updated with those obtained in the training phase. Without any doubts, the proposed hardware is extremely faster than the software implementation. Furthermore, it is much faster than the hardware engine proposed by Scott, Seth and Samal in [11]. However, it seems that our implementation requires almost twice the hardware are used to implement their architecture. Nevertheless, we do not have an exact record of the hardware are consumed in [11] as the authors did not provide nor include the hardware required to implement the fitness module for Michalewics's function [8].

References

1. Bade, S.L. and Hutchings, B.L., FPGA-Based Stochastic Neural Networks – Implementation, IEEE Workshop on FPGAs for Custom Computing Machines, Napa Ca, April 10–13, pp. 189–198, 1994.
2. Bland, I.M. and Megson, G. M., Implementing a generic systolic array for genetic algorithms. In Proc. 1st. On-Line Workshop on Soft Computing, pp 268–273, 1996.
3. Brown, B.D. and Card, H.C., Stochastic Neural Computation I: Computational Elements, IEEE Transactions on Computers, vol. 50, no. 9, pp. 891–905, September 2001.
4. Gaines, B.R., Stochastic Computing Systems, Advances in Information Systems Science, no. 2, pp. 37–172, 1969.
5. Liu, J., A general purpose hardware implementation of genetic algorithms, MSc. Thesis, University of North Carolina, 1993.
6. MathWorks, http://www.mathworks.com/, 2004.
7. Megson, G. M. Bland, I. M., Synthesis of a systolic array genetic algorithm. In Proc. 12th. International Parallel Processing Symposium, pp. 316–320, 1998.
8. Michalewics, Z., Genetic algorithms + data structures = evolution programs, Springer-Verlag, Berlin, Second Edition, 1994.
9. Nedjah, N. and Mourelle, L.M., Reconfigurable Hardware Architecture for Compact and Efficient Stochastic Neuron, Artificial Neural Nets Problem Solving Methods, Lecture Notes in Computer Science, vol. 2687, pp. 17–24, 2003.
10. Scott, S.D., Samal, A. and Seth, S., HGA: a hardware-based genetic algorithm, In Proc. ACM/SIGDA 3rd. International Symposium in Field-Programmable Gate Array, pp. 53–59, 1995.
11. Scott, S.D., Seth, S. and Samal, A., A hardware engine for genetic algorithms, Technical Report, UNL-CSE-97-001, University of Nebraska-Lincoln, July 1997.
12. Turton, B.H. and Arslan, T., A parallel genetic VLSI architecture for combinatorial real-time applications – disc scheduling, In Proc. IEE/IEEE International Conference on genetic Algorithms in Engineering Systems, pp. 88–93, 1994.
13. Xilinx, http://www.xilinx.com/, 2004.

4

Reconfigurable Computing and Parallelism for Implementing and Accelerating Evolutionary Algorithms

Miguel A. Vega Rodríguez[1], Juan A. Gómez Pulido[1], Juan M. Sánchez Pérez[1], José M. Granado Criado[1], and Manuel Rubio del Solar[2]

[1] Departamento de Informática,
Escuela Politécnica, Universidad de Extremadura,
Campus Universitario s/n, 10071 Cáceres, Spain
(`mavega,jangomez,sanperez,granado`)`@unex.es`, `http://arco.unex.es`
[2] Servicio de Informática, Universidad de Extremadura,
Avda. de Elvas s/n, Badajoz, Spain
`mrubio@unex.es`

Reconfigurable Computing is a technique for executing algorithms directly on the hardware in order to accelerate and increase their performance. Reconfigurable hardware consists of programmed FPGA chips for working as specific purpose coprocessors. The algorithms to be executed are programmed by means of description hardware languages and implemented in hardware using synthesis tools. Reconfigurable Computing is very useful for processing high computational cost algorithms because the algorithms implemented in a specific hardware get greater performance than if they are processed by a general purpose conventional processor. So Reconfigurable Computing and parallel techniques have been applied on a genetic algorithm for solving the salesman problem and on a parallel evolutionary algorithm for time series predictions. The hardware implementation of these two problems allows a wide set of tools and techniques to be shown. In both cases satisfactory experimental performances have been obtained.

4.1 Introduction

The reader will find two instances of hardware implementation of evolutionary algorithms in this chapter. Both instances provide distinct points of view on how to apply reconfigurable computing technology to increase algorithm efficiency. Two cases are considered: a genetic algorithm and a parallel

M.A. Vega Rodríguez et al.: *Reconfigurable Computing and Parallelism for Implementing and Accelerating Evolutionary Algorithms*, Studies in Computational Intelligence (SCI) **22**, 71–93 (2006)

evolutionary algorithm. As a result, many fields of reconfigurable hardware techniques are covered, such as modeling by means of high-level hardware description languages, implementation on different reconfigurable chips, hierarchical design, etc.

In Section 4.2 a general introductory view about Reconfigurable Computing and Field Programmable Gate Arrays (FPGAs) circuits, programming languages for algorithm modelling and implementing and the reconfigurable prototyped platforms used is offered. This will give us an idea about why this technology for the synthesis of the evolutionary algorithms is useful.

In Section 4.3 we perform a detailed study on the use of parallelism and FPGAs for implementing Genetic Algorithms (GAs). In particular, we do an experimental study using the Traveling Salesman Problem (TSP). After an overview on the TSP, we explain the GA used for solving it. Then, we study the hardware implementation of this algorithm, detailing 13 different hardware versions. Each new version improves the previous one, and many of these improvements are based on the use of parallelism techniques. In this section we also show and analyse the results found: Parallelism techniques that obtain better results, hardware/software comparisons, resource use, operation frequency, etc. We conclude stating FPGA implementation is better when the problem size increases or when better solutions (nearer to the optimum) must be found.

Finally, in Section 4.4 we show an application of Reconfigurable Computing for accelerating the execution of one of the steps of a parallel evolutionary algorithm. In the proposed algorithm the intermediate results evolve to find the local optimum. The algorithm has been created to increase the precision in the time series behaviour prediction. To do this, a set of processing units works in parallel mode to send its results to an evolutionary unit. This unit must determine an optimum value and generate a new input parameter sequence for the parallel units. The evolutionary unit has been implemented in the hardware in order to study its performance. We have found the algorithm execution is accelerated. This result encourages us to design, in the near future, a specific purpose processor for time series identification.

4.2 Reconfigurable Computing and FPGA

Reconfiguration of circuitry at runtime to suit the application at hand has created a promising paradigm of computing that blurs traditional frontiers between software and hardware. This powerful computing paradigm, named reconfigurable computing (or custom computing), is based on the use of field programmable logic devices, mainly field-programmable gate arrays (FPGAs) [15], incorporated in board-level reconfigurable systems. FPGAs have the benefits of the hardware is speed and the software is flexibility; also, they have a much more favourable price/performance ratio than ASICs (Application-Specific Integrated Circuits). For these reasons, FPGAs are a good alternative for many real applications in image and signal processing, multimedia, robotics, telecommunications, cryptography, networking and computation in general [16].

Furthermore, as reconfigurable computing is becoming an increasingly important computing paradigm, more and more tools are appearing in order to facilitate the FPGA programmability using higher-level HDLs (Hardware Description Languages). Along this line, several research projects have been developed to bring new high-level languages, most notably, SpecC [17], SystemC [18] or Ocapi-xl [19]. And notably, several companies are proposing their own high-level HDLs, such as Handel-C [20] or Forge [22].

The main advantage of all these new hardware description languages is their simplicity, where the hardware design can be defined and evaluated using a pseudo-C or pseudo-Java programming style. In this way, FPGA devices are making it possible for thousands of computer engineers to gain access to digital design technology more easily, obtaining better performance with similar software flexibility. In addition, ASIC engineers are now 'reconfiguring' themselves as FPGA engineers for economic reasons and adding to the growing legions of FPGA designers [27].

In our research works we use hardware resources such as prototyping boards. In Figure 4.1 our main platform is shown. The Celoxica RC1000 board [21] ships in a full length 64-bit PCI card a high-capacity FPGA (Xilinx Virtex V2000E-6/7/8-BG560), 4 independent asynchronous SRAM 512K x 32 banks, two PMC front panels (I/O via PC bracket and internal to PC), on board clock generator, 50 I/O connections routed from Virtex to 64 pin header memories, PCI bus direct to SelectMAP port, external JTAG connector and DMA. A diagram of these useful elements to be implemented in the designed systems is shown in Figure 4.2.

Fig. 4.1. The RC1000 prototyping platform for reconfigurable computing

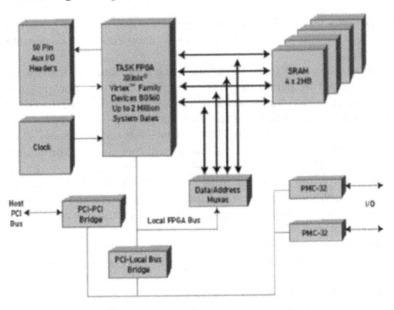

Fig. 4.2. The RC1000 scheme showing its main components

4.3 Implementing a Genetic Algorithm for Solving the TSP Using Parallelism and FPGAs

Genetic Algorithms (GAs) have been satisfactorily applied for solving many optimisation problems. Also, parallel techniques are useful to improve the efficiency of GAs [1]. However, a GA used for solving a large problem may be running during several days, even when it is run on a high-performance machine [2]. For this reason new methods that increase the velocity of the algorithms are needed. We believe that implementing a GA directly on hardware may increase its efficiency (decreasing the running time) in such a way that we can tackle more complex problems as well as perform more detailed research on the operation mode of the GA. In this work, we describe the hardware implementation, using FPGAs and parallelism techniques, of a GA for solving the TSP (Traveling Salesman Problem).

In other published studies related to the FPGA, the use of evolutionary computing may be found. In [7] a full review about these topics is performed, distinguishing between the research devoted to implementing exclusively the evaluation of fitness function on FPGAs, those that implement the full genetic or evolutionary algorithm, and those related with evolving hardware.

In this work the following steps have been taken: Using Visual C++ a genetic algorithm has been implemented for solving the TSP (see Figure 4.3). From this software implementation many experiments have been made in order to study the influence and importance of the different parameters (population size, crossover operator, etc.) on the quality of the solutions found

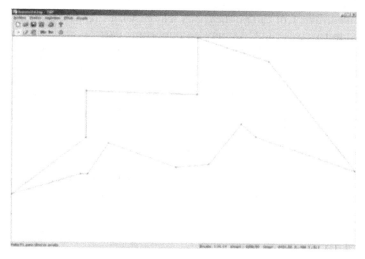

Fig. 4.3. Visual C++ application implementing the GA for the TSP

by the GA used. After obtaining the optimal values for the GA parameters, we have developed the hardware version of the GA (using a Virtex-E FPGA and the Handel-C language). Then, the hardware version has been optimised using internal arrays (FPGA resources instead of external memory banks) and parallelism techniques. In this way, the execution time has been considerably reduced. Finally we have analysed and compared the time measurements of the different implementations and we have concluded that the higher the work volume the more efficient/better the hardware version is, overcoming the software version if this work volume is sufficiently high.

4.3.1 The TSP

In the TSP we have a set of nodes, which represents cities or, simply, places. Starting from a node, we have to visit all the nodes, visiting each node exactly once, and stopping at the initial node. Each arc/union of two nodes is characterized by a C_{ij} value, which indicates the distance or cost of going from the i node to the j node. To solve the problem, it is necessary to find the order in which the nodes must be visited with the goal that the travelled distance/cost is the minimum. In this work, we use the more classical and popular TSP version: The symmetric TSP.

Although the TSP seems an easy problem, for N nodes, the number of possible tours is (N-1)!/2, and consequently, the total time to find all the tours is proportional to N!, quite a high number [10]. In fact, within the complexity theory, the TSP is considered as a NP-complete combinatorial problem, making it necessary to use a heuristic search. In publications may be found many alternatives for solving this problem. In particular, in [6] there is a detailed review of the publications related with this problem. Among the possible alternatives the GAs play an important part.

The TSP has a great number of practical applications: It is the bases of many applications of transportation, logistics and delivery of merchandise. It is also used in the scheduling of machines to drill holes in circuit boards or other objects, in the construction of radiation hybrid maps as part of the work in genome sequencing, in the design of fiber optical networks, etc. [8].

4.3.2 The Genetic Algorithm Used

We use a genetic algorithm based on the classical GAs. In our case, the first step of the GA consists of creating randomly the initial population that has 197 individuals. These individuals will be ordered according to the fitness function. Then, the following actions are made in each iteration of the GA (Figure 4.4):

- Three members of the population are randomly selected.
- These three members/parents are crossed among them to generate three new offspring.
- The population individuals suffer mutations with a 1% probability.
- The three new offspring are added to the population and ordered according to their fitness function. At this time the population is at its highest, 200 individuals.
- The three worst adapted members are removed, and therefore, the population has again 197 individuals.

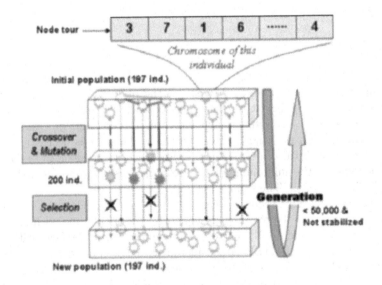

Fig. 4.4. Genetic algorithm used

In our case, the algorithm stopping condition is reached when either the maximum iteration number (50,000) is finished or when the population is stabilized. We consider the population to be stabilized when the best adapted individual is not modified during 1,000 iterations.

It is important to emphasize that the election of all these values (population of 200 individuals, crossover among 3 parents, mutation probability of 1%, maximum number of iterations 50,000, stabilization at 1,000 iterations, etc.) has been made after performing many experiments. These values have been chosen because they lead to the best results.

An individual consists of a cyclical tour, for this reason, we have used the path representation [6], which is considered to be the most natural representation of a tour. Furthermore, this is the most used representation for the TSP with GAs, since it is the most intuitive and has also lead to the best results [6]. The crossover operator used is the ER (genetic Edge Recombination crossover). This operator was developed by Whitley et al. [13],[14], and according to [6], it is one of the crossover operators better adapted to the TSP. The mutation operator used in this work is the IVM (InVersion Mutation) [4],[5], which is the most suitable for the TSP with the ER crossover operator [6]. The IVM operator is also called Cut-Inverse Mutation Operator [3] in other works. Finally, the fitness function selected is the sum of all the weights (distances or costs, C_{ij}) of all the arcs/unions in the cyclical tour. The best adapted individuals will have the minimum value fitness function.

4.3.3 Hardware Implementation Using Parallelism and FPGAs

For the hardware implementation of the genetic algorithm a Xilinx Virtex-E FPGA [22] has been used. Concretely, the XCV2000E-8 FPGA, included in the Celoxica RC1000 board [21]. The hardware description has been performed by means of Handel-C language and its DK1 design suite [20]. Afterwards, we have used the Xilinx ISE tool [22] in order to convert the EDIF files, obtained thanks to DK1, into the corresponding. BIT files (files that contain the bitstream for configuring the FPGA).

For generating the random numbers in the GA (i.e., for creating randomly the initial population, controlling the mutation, etc.) a Linear Feedback Shift Register (LFSR) of 16 bits has been used. As a starting point we have designed, using Handel-C, a GA similar to that of the software version in Visual C++, without adding parallelism and making an extensive use of the RC1000 board memory banks. In successive hardware versions of the genetic algorithm, improvements to the hardware programming have been added in order to compare their results. We summarize the modifications performed in each of the different hardware versions of the GA:

- Version 1 (V1): Original version, similar to the software version in Visual C++.

- Version 2 (V2): Modification of the individual creation. With the new method, it is not necessary to go over the whole node array (to check if the node already exists within the tour/individual) each time a node of the new individual is generated. The new method uses a standard individual/tour (with all the nodes in order: 1-2-3-4-), and then, to create the new individual it disorders its nodes: for each node it selects another second node randomly and interchanges them. The $Create\_Individual()$ method is modified, allowing the access in parallel to different nodes of an individual (use of parallelism).

- Version 3 (V3): Parallelism statements are introduced throughtout the code (**par** statements in Handel-C). In Handel-C, all the statements included in a **par** code block will operate in parallel.

- Version 4 (V4): Creation of the internal RAMs (FPGA resources instead of external memory banks) for DistanceArray and MappingArray, notably saving in each iteration many accesses to the board memory banks (they are slower).

- Version 5 (V5): Maximal reduction of the sizes in bits of the variables and structures used in hardware. The internal RAM Individual is created in the $Create\_Individual()$ method.

- Version 6 (V6): Parallelism is added to the Sorting() method, and in all the code where DistanceArray and MappingArray are used. Now, DistanceArray and MappingArray are different internal RAMs and we can access them at the same time. Before Version 4, it was not possible because they were stored in the same memory bank (only one access to the memory bank at a time).

- Version 7 (V7): Improvement in the ER crossover algorithm and in the neighbour map [6]. This map is implemented as an internal RAM (EdgeMap structure). In the $Create\_Offspring()$ method the FlagArray structure is added as an internal array, and the CandidateArray structure is implemented as an internal array instead of internal RAM to use it with parallelism (multiple accesses in parallel). An internal array, like an internal RAM, uses FPGA resources instead of memory banks, providing faster access. Furthermore, an internal array allows multiple accesses to different array positions in parallel (an internal RAM only allows one access to one array position at the same time). An internal array consumes more FPGA resources than an internal RAM.

- Version 8 (V8): Improvement in the mutation. In the $Mutate\_Individual()$ method, the auxiliary structures Individual and $Aux\_Population$ are incorporated by means of internal arrays (FPGA resources). In this way, we can apply parallelism to these structures (multiple accesses in parallel).

- Version 9 (V9): The $Sorting()$ method is removed, and it is substituted by a method of insertion in order ($Insert\_Order()$). This implies that the process of initial creation of the population (it is the only one that uses the sorting) may be placed outside the FPGA. All this also involves the removal of the methods $Extend\_Population()$,

Create_Initial_Population() and *Create_Individual*(), because they are now not necessary.

- Version 10 (V10): Data replication. The distance triangular matrix is stored in the four memory banks of the board to access memory four times at the same time when the individual distances are computed. That is, making the memory accesses in parallel, four times quicker.
- Version 11 (V11): Data compaction. The node data is 7-bit in size. They are stored from four in four in each memory location (32 bits). So, four nodes are read in only one access. This allows us to include more parallelism. In the previous versions, to keep things simple, each memory location only stored one node.
- Version 12 (V12): This version is the same as version 8, but with Stabilization at 16,000 iterations. In this way, we can check if the efficiency of the hardware implementation improves when we increase the stabilization.
- Version 13 (V13): This version is the same as version 11, but with Stabilization at 16,383 iterations (maximum value with 14 bits). Furthermore, we have removed the *Annul_Individual*() method, because it is now not necessary.

As we can observe, many of the improvements included in the successive hardware versions are based on the use of parallelism techniques. Concretely, the hardware versions 2, 3, 6, 7, 8, 10 and 11 include improvements based on the use of parallelism.

4.3.4 Results

In order to perform our experiments we have used the TSPLIB library [9]. TSPLIB is a library of sample instances (benchmarks) for the TSP collected from multiple sources. These instances cover the different TSP types (symmetric, asymmetric, Hamiltonian cycles,etc.). All the files belonging to this library are in a specific format (see Figure 4.5(a) for an example).

In conclusion, the TSPLIB library contains a great number of TSP instances from around the world in a normalized format, but also it contains the best solution known for each instance up to date (Figure 4.5(b)). This fact converts this library into a reference for everyone that works with the TSP.

Table 4.1 shows the ratio between the average execution time of the different hardware versions (XCV2000E FPGA of the RC1000 board) and the corresponding software versions (1.7-GHz Pentium-IV with 768 MB of RAM). These data are shown for multiple instances/problems of the TSPLIB library. Each software implementation has the same algorithmic improvements as its corresponding hardware implementation. In this table have been obtained after 10 executions of each experiment. For every instance (TSP problem), both implementations (hardware and software) provide solutions of identical quality (in fact, due to the pseudo-random number generation both implementations provide identical solutions). Although we do not show these solutions

Fig. 4.5. (a) Example of instance. (b) Optimal solution for this instance

Table 4.1. Ratio between the HW/SW average execution times

Instance	V1	V2	V3	V4	V5	V6	V7
Burma14	38.627	36.918	36.773	25.003	24.444	22.899	22.396
Ulysses16	33.277	31.085	30.902	20.126	19.546	18.241	17.723
Ulysses22	23.358	24.020	23.893	15.589	14.962	14.157	13.594
Gr24	23.914	22.060	21.928	14.318	13.685	12.992	12.423
Bayg29	19.252	18.897	18.775	12.524	11.889	11.396	10.827
Bays29	17.558	18.255	18.133	11.858	11.220	10.727	10.155
Gr48	10.801	10.870	10.780	7.435	6.863	6.690	6.182
Eil51	13.774	9.685	9.597	6.670	6.111	5.961	5.465
Berlin52	8.905	9.340	9.255	6.256	5.702	5.557	5.065
Brazil58	7.402	8.905	8.826	6.310	5.786	5.676	5.211
St70	6.622	7.075	7.009	5.229	4.761	4.693	4.281
Pr76	5.825	5.985	5.928	4.342	3.906	3.852	3.468
Rat99	4.070	3.759	3.721	2.838	2.507	2.482	2.193
Rd100	3.749	4.124	4.077	3.134	2.787	2.761	2.457

Instance	V8	V9	V10	V11	V12	V13
Burma14	22.383	19.382	19.317	19.204	14.024	14.919
Ulysses16	17.704	17.258	17.198	17.078	12.160	12.672
Ulysses22	13.581	11.662	11.602	11.474	8.640	8.723
Gr24	12.416	12.599	12.538	12.403	7.539	7.861
Bayg29	10.815	9.769	9.706	9.573	5.861	6.574
Bays29	10.143	10.032	9.970	9.837	5.832	6.285
Gr48	6.175	7.050	7.001	6.870	2.961	3.489
Eil51	5.458	5.354	5.305	5.180	2.914	3.216
Berlin52	5.059	4.692	4.643	4.523	2.876	2.975
Brazil58	5.206	4.219	4.175	4.060	2.460	2.598
St70	4.277	3.712	3.674	3.562	2.055	2.244
Pr76	3.464	3.385	3.352	3.245	1.567	2.137
Rat99	2.191	2.280	2.257	2.172	1.312	1.747
Rd100	2.455	2.297	2.269	2.183	1.424	1.493

for space reasons, we indicate that in all the cases we have found the optimal solution or a near solution. Note that the instance names include a suffix that represents the instance size/difficulty, that is, its number of nodes.

In order to simplify easier the interpretation of the data in Table 4.1, Table 4.2 indicates the differences in the HW-time/SW-time ratio between a present version and the previous one. In this table, the last row shows the average of the differences for each column. Using these averages we can

Table 4.2. Differences in the HW-time/SW-time ratio between a version and the previous one

Instance	V2-V1	V3-V2	V4-V3	V5-V4	V6-V5	V7-V6
Burma14	1.709	0.145	11.77	0.559	1.545	0.503
Ulysses16	2.192	0.183	10.776	0.58	1.305	0.518
Ulysses22	-0.662	0.127	8.304	0.627	0.805	0.563
Gr24	1.854	0.132	7.61	0.633	0.693	0.569
Bayg29	0.355	0.122	6.251	0.635	0.493	0.569
Bays29	-0.697	0.122	6.275	0.638	0.493	0.572
Gr48	-0.069	0.09	3.345	0.572	0.173	0.508
Eil51	4.089	0.088	2.927	0.559	0.15	0.496
Berlin52	-0.435	0.085	2.999	0.554	0.145	0.492
Brazil58	-1.503	0.079	2.516	0.524	0.11	0.465
St70	-0.453	0.066	1.78	0.468	0.068	0.412
Pr76	-0.16	0.057	1.586	0.436	0.054	0.384
Rat99	0.311	0.038	0.883	0.331	0.025	0.289
Rd100	-0.375	0.047	0.943	0.347	0.026	0.304
Average	0.440	0.099	4.855	0.533	0.435	0.475

Instance	V8-V7	V9-V8	V10-V9	V11-V10	V12-V11	V13-V12
Burma14	0.013	3.001	0.065	0.113	5.18	-0.895
Ulysses16	0.019	0.446	0.06	0.12	4.918	-0.512
Ulysses22	0.013	1.919	0.06	0.128	2.834	-0.083
Gr24	0.007	-0.183	0.061	0.135	4.864	-0.322
Bayg29	0.012	1.046	0.063	0.133	3.712	-0.713
Bays29	0.012	0.111	0.062	0.133	4.005	-0.453
Gr48	0.007	-0.875	0.049	0.131	3.909	-0.528
Eil51	0.007	0.104	0.049	0.125	2.266	-0.302
Berlin52	0.006	0.367	0.049	0.12	1.647	-0.099
Brazil58	0.005	0.987	0.044	0.115	1.6	-0.138
St70	0.004	0.565	0.038	0.112	1.507	-0.189
Pr76	0.004	0.079	0.033	0.107	1.678	-0.57
Rat99	0.002	-0.089	0.023	0.085	0.86	-0.435
Rd100	0.002	0.158	0.028	0.086	0.759	-0.069
Average	0.008	0.545	0.049	0.117	2.839	-0.379

estimate the overall improvement of a version with respect to the previous one.

From Tables 4.1 and 4.2, we conclude that the advantage of the software version decreases when the problem size (number of nodes) increases, that is, when more time is necessary to solve the TSP problem. On the other hand, when hardware versions are improved (by the use of parallelism, reduction of accesses to memory banks by internal arrays, etc.) the HW-time/SW-time ratio decreases; putting the hardware's performance closer and closer to that of the software. Assuming an improvement is significant when it surpasses the average 0.5 in the last row of Table 4.2, the more significant changes/improvements in this ratio are produced in versions 4, 5, 9, 12 and 13 (comparing version 13 with respect to version 11, from where version 13 comes). The improvement of version 4 indicates that the use of internal resources of the FPGA for data storing allows us to reduce the number of accesses to memory banks, improving the hardware implementation efficiency. Version 5 shows us that the use of the exact number of bits required for each datum is important in the hardware implementations. In version 9, where the individual sorting is changed by an insertion in order, an important time improvement is obtained in the hardware version, which is seen as more penalized than the software version in accessing data. This improvement will be more significant while the problem size is smaller, because it will have more weight in the algorithm. The improvements in versions 12 and 13 display that increasing the Stabilization value (that is, the genetic algorithm must do more iterations), a point at which the software version is slower than the hardware version can be reached, because the more iterations performed the more advantages for the FPGA implementation.

From the point of view of parallelism, these are the hardware versions that include improvements based on the use of parallelism, and the parallelism techniques applied:

- V2: Data parallelism due to multiple accesses to different array positions in parallel.
- V3: Functional parallelism [11], performing several statements at the same time.
- V6: Data parallelism due to the access of different arrays in parallel.
- V7 and V8: Data parallelism due to both multiple accesses to different array positions and several accesses to different arrays at the same time.
- V10: Data parallelism due to the replication of data. For very large arrays (for example, the distance triangular matrix), we have to store them in the memory banks of the board because there are not enough FPGA resources for implementing them using internal RAMs or internal arrays. In these cases, we have created several copies of the same array in different memory banks. Therefore, we can perform several accesses to the same array in parallel, even though it is in memory banks.
- V11: Data parallelism due to data compactation. Now, each memory bank location (32 bits) does not store one node (7 bits) of an individual, but it stores four compacted nodes in the same memory word. In this way, in only one access we can read four times more data, and thus, we can apply more parallelism.

In summary, we have used very different parallelism techniques, applying the two possible kinds of parallelism [11]: functional parallelism and data parallelism. In this case, data parallelism obtains greater improvements in the hardware implementations. On the one hand, in our GA's code there are many data accesses, and on the other hand, this high number of accesses limits the parallelization of other code portions. In particular, from table 4.2 we conclude that the parallelism techniques that obtain better results are the ones applied to versions 7, 2 and 6, in this order. Version 8 generates the worst improvements because this version optimises the mutation, and even though this is a good optimisation, it is seldon used (mutation probability is 1%).

Table 4.3 shows the comparisons of FPGA occupation (resource use) and maximum work frequency for the different hardware versions [12]. The resource use of the FPGA increases while we apply more and more improvements to the hardware versions, because the improvements imply the parallelism exploitation, the use of internal arrays for data storing, etc. We highlight versions 5 and 9. In version 5, due to the fact that we have reduced the size in bits of the variables and data structures, the FPGA occupation has been decreased by more than 50%. In version 9, the substitution of the sorting by the insertion in order also releases FPGA resources. Finally, remember that version 12

Table 4.3. Resource use, frequencies and periods for the different hardware versions

Version	Resource use (slices)	FPGA occupation (% of 19,200)	Maximum frequency (MHz)	Minimum period (ns)
V1	4696	24%	14.641	68.301
V2	6316	32%	15.108	66.188
V3	6294	32%	15.588	64.152
V4	5575	29%	14.426	69.321
V5	**2728**	**14%**	**12.793**	**78.167**
V6	4007	20%	11.157	89.629
V7	7821	40%	12.194	82.007
V8	10411	54%	13.008	76.874
V9	**8984**	**46%**	**13.927**	**71.805**
V10	11954	62%	11.881	84.167
V11	16858	87%	13.224	75.620
V12	10411	54%	13.008	76.874
V13	16796	87%	13.807	72.428

is identical to version 8, but with Stabilization at 16,000 iterations (for this reason, both versions have the same results).

As for the frequency (and the period), this is practically stable in all the hardware versions (varying slightly between 11.157 MHz and 15.588 MHz).

4.3.5 Conclusions

In this work we have performed a study on the implementation of GAs by means of parallelism and FPGAs, using the TSP as a case study. Figure 4.6 summarizes graphically some of the more important conclusions.

The evolution of the HW/SW ratio, in the 13 versions, indicates clearly that, for problems whose size is greater than 100 nodes, the FPGA implementation will become more efficient than the software implementation, increasing this difference when the problem is bigger or more complex (more time for solving it). The evolution in the 13 versions also displays that each new version improves the previous one.

It is clear that after fixing the best parameters and the best GA for the problem, the quality of the obtained solutions will depend on the time we employ in finding the solution, that is, the number of iterations used for obtaining the solution. In our case, it will depend on the Stabilization parameter. As it can be seen in this Figure 4.6 and in Table 4.1, the higher the stabilization is, the more efficient the hardware (FPGA) version is with respect to the software version. In conclusion, the hardware implementation also gathers strength in this aspect, since it will be more necessary when more work has to be done.

From the point of view of the parallelism, we have studied very different parallelism techniques in order to improve the efficiency of the FPGA implementation of GAs. From this work, using the TSP, we conclude that we obtain

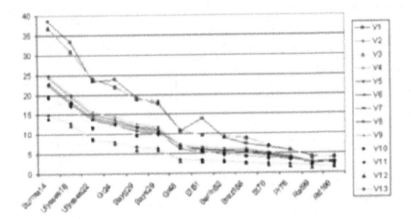

Fig. 4.6. Graphical representation of the ratio between the HW/SW average execution times

greater improvements using the data parallelism. Particularly, the data parallelism techniques that include multiple accesses to different array positions or several accesses to different arrays at the same time.

Finally, another important conclusion is that FPGAs do not adapt in the same way to all the GAs. In the TSP case, the fitness function implies a great number of memory accesses for reading both the individuals and the node list of each individual (nodes in each tour), or even the distances between different nodes. Such a great number of memory accesses penalizes the efficiency of the hardware version in relation to the software one. For this reason, we believe that other GAs, whose fitness functions have fewer memory accesses and more computation necessities, would be more suitable for FPGA implementation. In the TSP case, the computation is minimal, because the fitness function only has to add the different distances between the nodes in the tour/individual.

4.4 Hardware Acceleration of a Parallel Evolutionary Algorithm

In this section a second example of hardware implementation of a parallel evolutionary algorithm is shown. Now, this case is useful to illustrate the synthesis from another point of view, because the algorithm has been modeled from schematics, logic devices and finite state machines. As in the TSP case, a study about the hardware/software improvement has been performed.

4.4.1 The Problem: Modeling and Predicting Time Series

In many engineering fields it is necessary to make use of mathematical models for studying the behaviour of dynamic systems whose mathematical description is not available 'a priori'. One interesting type of these systems is the Time Series (TS); they are used to describe systems in many fields: meteorology, economy, physics, etc. We employ System Identification (SI) techniques [23] in order to obtain the TS model. We consider a TS as the description of a simple dynamic system, by means of a sampled signal with period T that is modelled with an ARMAX [24] parametric description. Basically the identification consists of determining the ARMAX model parameters a_i from measured samples $y(k_i)$. Then it is possible to compute the estimated signal $y_e(k)$ (Equation 4.1: estimated value of the TS at k time) and compare it with the real signal y(k), computing the generated error at k time.

$$y_e(k) = [-a_1 y(k-1) - \ldots - a_{na} y(k - na)] \qquad (4.1)$$

The recursive estimation updates a_i in each time step k, thus modelling the system. The more sampled data processed, the more precision for the model, because it has more information about the system behaviour history. We consider SI performed by the Recursive Least Squares (RLS) with forgetting factor (λ) algorithm [24], in order to evaluate y_e. This algorithm is specified by the constant λ, the initial values and the observed samples y(k). There is not any fixed value for λ, even when the value is between 0.97 and 0.995. The cost function F (see Equation 4.2, where SN is the sample number) is defined as the value to minimize in order to obtain the best precision.

$$F = \sum_{k=k_0}^{k=k_0+SN-1} |y_e(k) - y(k)| \qquad (4.2)$$

When SI techniques are used, the model is generated 'a posteriori' by means of the measured data. However we are interested in the system behaviour prediction in running time, that is, while the system is working and its data are being observed. So, it would be interesting to generate models in running time in such a way that a processor may simulate the system next behaviour. But the precision problem may appear when a system model is generated in sample time: If the system response changes quickly, then the sample frequency must be high to avoid the key data loss in the system behaviour description. If the system is complex and its simulation from the model to be found must be very trustworthy, then the required precision must be very high and this implies a great computational cost. Sometimes the hardware resources do not allow the computational cost in the model generation and processing to be lower than the sample period. On the other hand, the precision (minimal F) is due to several causes, mainly to the forgetting factor λ (Figure 4.7). Frequently this value is critical for model precision. Other sources can also have less influence (model dimensions, etc.), but they are considered as problem definitions, not parameters to be optimized. For example, we can fix an adequate value of the models size in na $=15$ [25].

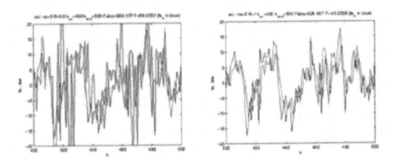

Fig. 4.7. RLS identification of benchmark ball with different values for λ. In the left case ($\lambda=0.6$) the produced error is greater than in the right case ($\lambda = 1$)

4.4.2 A Parallel Evolutionary Heuristic

In order to find the optimum value of λ, we propose a parallel algorithm that is partially inspired by the concept of artificial evolution [26]. In our algorithm, named PERLS (Parallel Evolutionary Recursive Least Squares), the optimization parameter λ is evolved to predict new situations during the successive phases of the process. In other words, λ evolves at the same time which improves the cost function performance. PERLS considers a λ value as a state. Starting on an initial λ value (λ_c) and an initial R value (the interval of generation where λ_c is in the middle), a set of λ values is generated covering uniformly the interval R, with equal distances between them. The λ values generated are equal to the number of parallel processing units (PUN). Each phase of PERLS is an identification loop that considers a given number of sample times (PHS) and the corresponding λ value. We use the nomenclature shown in Table 4.4.

In each phase, R is reduced dividing itself by the RED factor (the interval limits are moved so the center of interval corresponds with the optimal λ value found in the previous phase), in such a way that the generated set of λ will

Table 4.4. PERLS nomenclature: the more important algorithm's parameters

R	Generation interval	**PUN**	Number of parallel processing units
λ_c	λ central in R	**TSN**	Total number of samples
PHS	Phase samples	**RED**	Reduction factor of R
PHN	Number of phases		

be closer and closer to the previous optimum found. The new set of generated λ values covers uniformly the new R. In each processing unit, during each phase, the cost function F is computed (the accumulated error of the samples that constitutes each phase). At the end of each phase, the best λ is chosen. This is the corresponding value to the lower F. From this λ, new values are generated in a more reduced (new R) interval (see Figure 4.8). The goal is that the identifications performed by the processing units will converge to optimum λ parameters when a given stop criteria is achieved. In Figure 4.9 we can see the parallel architecture what would help us to better understand the PERLS performance.

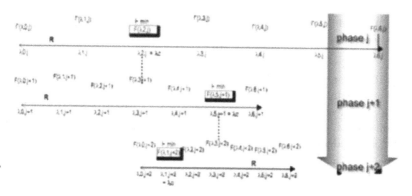

Fig. 4.8. PERLS behaviour. Identification uses different λ values in each phase performed by each processing unit. All the λ values in the same phase running in the processing units are generated in the R interval from the previous phase optimum λ found, corresponding with the smallest computed F. Thus, in the phase j the minimum F determines the optimum λ (in this example, $\lambda_{4,j}$), and in the phase j+1, R is reduced and the new λ_c is the optimum found in the phase

4.4.3 Design of a Reconfigurable Processor

In our proposal, a time series prediction designed board has got an specific processor in which the PERLS algorithm is implemented, together with D/A and A/D converters, a general controller and a system clock. The time series is sampled and digitalized by means of a data acquisition board and then it is sent to the processor. This (the processor) generates the estimated output signal and the ARMAX model to be monitored. PERLS is implemented by a set of parallel process units (PUs). These units perform identifications in each phase, each one with its corresponding value of the λ parameter. When a phase ends and before the next sample time where a new phase starts, each PU sends λ and F to the Evolutionary Central Unit (ECU). Then the ECU generates a new range of λ values for all PUs in the new phase. The implementation

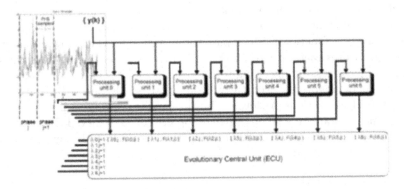

Fig. 4.9. PERLS architecture. A set of processing units generates the next set of λ values to be computed by RLS algorithm

of a PU is based on the RLS algorithm, while the ECU implementation is determinated by the evolutionary strategy.

We have designed, implemented and tested the ECU unit so as to verify the suitability of applying the reconfigurable computation for the PERLS evolutionary algorithm. The designed ECU (Figure 4.10(a)) is adapted to PERLS with the following settings: PUN = 5 and RED = 2. At this point we consider the time series values to be digitalized. ECU consists of the following I/O signals:

- INIT is an input signal for initiating the internal controller based on a finite state machine.
- SET is an input signal for controlling the cyclic sequence of the phases.
- F0[16] ... F4[16] are the cost function values coming from the parallel processing units, digitalized with a wide of 16 bits.
- FF0[16] ... FF4[16] are the forgetting-factor parameter values corresponding to the previous cost functions, also digitalized with a wide of 16 bits.
- NPHASE[4] returns the phase number in processing.
- FMIN[16] returns the found optimal cost function.
- FF0S[16] ... FF4S[16] are the forgetting-factor new values returned in order to be used in the next phase by the parallel processing units.

The ECU consists of a set of modules (Figure 4.10(b)) with capacity for comparing the cost function F values, arithmetic generation of new λ values, etc. Two ROM memories store the successive generation intervals. So, the successive values of R/2 and R/4 are read from the ROM-R2 and ROM-R4 memories, respectively. The output adders generate the new value range from the read values of ROM memories and the optimal value found in the comparison module. The process is controlled by a finite state machine (Figure 4.10(c)) that, on the one hand, sets PERLS through the INIT signal, and on the other hand, it allows ECU to work only during the phase transitions, that is, between the two sample times that indicate the end of one phase and the beginning of the next. ECU starts to work by means of SET signal coming from the PERLS controller. The comparison module (Figure 4.10(d)) is composed of 5 sequential comparators of input buses that drive the F data. With the goal that the value reference corresponding to each F will not be lost, the buses with F are routed together through this sequential circuit. Experimentally it has been tested that this circuit is faster than that obtained by synthetizing the programmed circuit using the hardware description language VHDL.

4.4.4 Experimental Results

The ECU has been simulated and synthetized using Xilinx Foundation 4.2i [22], generating the same values that those found by means a probe code writen in C. Table 4.5 shows the different delay measurements corresponding with the ECU for two different FPGAs, according to a time analysis performed after the layout generation.

Table 4.5. Timing results of the Evolutionary Central Unit

FPGA	Post layout timing report
xc4062xla07	Minimum period: 5.961ns (Maximum frequency: 167.75MHz) Maximum combinational path delay: 34.696ns Maximum net delay: 18.876ns
xcv1600e8fg860	Minimum period: 5.224ns (Maximum frequency: 191.42MHz) Maximum combinational path delay: 19.332ns Maximum net delay: 9.954ns
xcv2000e8bg560	Minimum period: 5.520ns (Maximum frequency: 181.16MHz) Maximum combinational path delay: 21.737ns Maximum net delay: 9.287ns

We have measured the CPU cycle number that uses the compiled program from the tested C code; the time taken by this program and that of the FPGAs is then compared. As the measured cycles depend on the used processor and the compiler type, we analyze carefully the generated machine code by the compiler. To do that we have consider several methods:

- For an Intel architecture, it is possible to read the CPU ticks counter register using the RTCS instruction, finding the real CPU time.
- An assembler instruction study may be done, knowing the cycle number spent by each instruction. This is a tedious method that approximates the user time, but it does not take into account the same characteristics of processor architecture.
- Software tools may be used as gprof (it does not permit very low resolutions of time measures), Vtune, Quantyfy, etc.

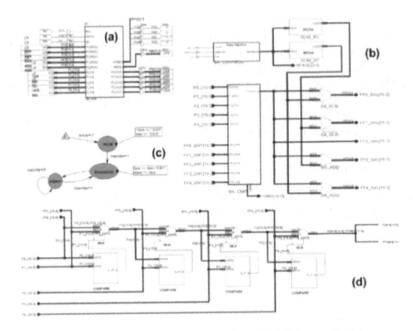

Fig. 4.10. PERLS implementation in reconfigurable hardware. In (a) we can see the symbol of the top-level schematic of the unit, with the I/O pad connections to FPGA pins. In (b) we can see the different elements inside the top-level symbol: two rom memories containing the successive sizes of the λ generation intervals, bus adders for generating the new set of λ values in the new interval, one control unit to drive the process and one comparator unit to select the smallest F found. The adders generate the new set of λ values from the optimum found (λ_c) in the comparator unit. In (c) the schema of the finite state machine that drives the process is shown. Finally, in (d) the buses and comparator elements that select the smallest F, so that from it the optimum λ (λ_c) of this phase can be determined, are shown

We have selected the first method, compiling the code using MS Visual C++ 6 with the Intel C++ 5.0 compiler option, under Windows XP, with a Pentium IV of 1,5 GHz (T = 0,66 ns), inserting RDTSC counters. In the execution we obtain 411 CPU cycles, that are 274 ns. It must be observed that this time includes all O.S. processes in execution in the CPU at the same time as the user program. Several thousands of measures for obtaining the medium value of the user time have been made, substracting the corresponding cycles to the O.S. processes. However the architecture details for reducing (pipelining) or increasing (caches) this time has not been able to be considered. So, for taking into account the processor architecture cycles and impact, we use Intel Vtune. The static and running time analysis give us the used CPU cycles, counting miss penalizations by memory accesses, etc.: 138 instructions corresponding to 1934 CPU cycles, that is 1290 ns.

4.4.5 Conclusion

The changes made to a program that performs exactly the same operations as the ECU circuit produces faster execution times (over 200ns) than those obtained by the FPGA implementation (about 19ns as we can see in Table 4.5, with FPGAs depending on the year of the Intel processor used). This program and this circuit correspond to the generation module in only one phase. So, assuming that the RLS execution module will also be competitive on an FPGA (highly probable), the sum of the performances of both modules is multiplied by all the successive phases of the PERLS algorithm, predicting, as a consequence, a satisfactory final performance. Also, we note that the synthesis technology used (Xilinx Foundation 4) is less optimized than the last tools (Xilinx ISE 8).

4.5 Summary

In this chapter two works, that implement evolutionary algorithms on Reconfigurable Hardware with different techniques and tools, have been exposed. In both cases parallel techniques have been used. The reason for using FPGA chips for the algorithm execution is based on the increasing performance they provide versus the general purpose processors. In the first work, a genetic algorithm hardware implementation for solving the salesman problem is presented. Parallel techniques for improving the hardware implementation have been used. Good execution time results have been found. In the second work, the implementation of a phase of a parallel evolutionary algorithm is presented. This has been made to increase the time series behaviour prediction precision. The implemented step has produced better execution times than those produced by a contemporary processor using FPGA. Both experiments promote the suitability of designing specific purpose processors on reconfigurable circuits, overall when relieving the elevated computational work load associated with some algorithms. This allows, on the one hand, to accelerate the algorithm, and on the other hand, to lighten the generic processor work load.

With regard to future work we have thought about using more advanced prototyping boards that extend the FPGA capabilities, so algorithms of high computational cost can be tackled. We also want to implement on hardware parallel evolutionary algorithms that deal with optimization problems in the networks field.

Acknowledgements

This work has been developed thanks to TRACER project (TIC2002-04498-C05-01, Ministerio de Ciencia y Tecnología, Spain, 2002-2005) working on it the following Spanish universities: Universidad de Extremadura, Universidad de Málaga, Universidad Politécnica de Cataluña, Universidad de La Laguna and Universidad Carlos III de Madrid.

References

1. E. Alba, M. Tomassini, "Parallelism and evolutionary algorithms", *IEEE Transactions on Evolutionary Computation*, 6, IEEE Press, 2002, pp. 443-462.
2. C. Aporntewan, P. Chongstitvatana, "A Hardware Implementation of the Compact Genetic Algorithm", *IEEE Congress on Evolutionary Computation*, IEEE 2001, pp. 624-629.
3. W. Banzhaf, "The 'Molecular' Traveling Salesman", *Biological Cybernetics* 64, Springer-Verlag, 1990, pp. 7-14.
4. D.B. Fogel, "A Parallel Processing Approach to a Multiple Traveling Salesman Problem Using Evolutionary Programming", *Proc. 4th Annual Parallel Processing Symposium*, Fullerton, CA, 1990, pp. 318-326.
5. D.B. Fogel, "Applying Evolutionary Programming to Selected Traveling Salesman Problems", *Cybernetics and Systems* 24, Taylor-Francis, 1993, pp. 27-36.
6. P. Larrañaga, C.M.H. Kuijpers, R. Murga, I. Inza, S. Dizdarevic, "Genetic Algorithms for the Traveling Salesman Problem: A Review of Representations and Operators", *Artificial Intelligence Review* 13, Springer, 1999, pp. 129-170.
7. P. Martin, "A Hardware Implementation of a Genetic Programming System using FPGAs and Handel-C", *Genetic Programming and Evolvable Machines* 2(4), Springer, 2001, pp. 317-343.
8. E. Lawler, J. Karel, A. Rinnooy and D. Shmoys, *The Traveling Salesman Problem, A Guided Tour of Combinatorial Optimization*, Wiley, 1985.
9. Jünger, M., G. Reinelt and G. Rinaldi, "The Traveling Salesman Problem", *Annotated Bibliographies in Combinatorial Optimization*, M. Dell' Amico, F. Maffioli, S. Martello (eds.), Wiley, 1997, pp. 199-221.
10. E. Rich, K. Knight, *Artificial Intelligence*, McGraw-Hill, 1991.
11. D. Sima, T. Fountain, P. Kacsuk, *Advanced Computer Architecture: A Design Space Approach*, Addison-Wesley, 1998.
12. M.A. Vega, R. Gutiérrez, J.M. Ávila, J.M. Sánchez, J.A. Gómez, "Genetic Algorithms Using Parallelism and FPGAs: The TSP as Case Study", *Proc. 2005 International Conference on Parallel Processing*, IEEE Computer Society, 2005, pp. 573-579.
13. D. Whitley, T. Starkweather, D. Fuquay, "Scheduling Problems and Traveling Salesman: The Genetic Edge Recombination Operator", *Proc. 3rd International Conference on Genetic Algorithms*, Los Altos, CA, USA, 1989, pp. 133-140.
14. D. Whitley, T. Starkweather, D. Shaner, "The Traveling Salesman and Sequence Scheduling: Quality Solutions Using Genetic Edge Recombination", *Handbook of Genetic Algorithms*, Van Nostrand Reinhold, New York, USA, 1991, pp. 350-372.
15. B. Zeidman, *Designing with FPGAs and CPLDs*, CMP Books, 2002.
16. M.A. Vega, J.M. Sánchez, J.A. Gómez, "Guest Editors' Introduction – Special Issue on FPGAs: Applications and Designs", *Microprocessors and Microsystems*, 28(5-6), Elsevier Science, 2004, pp. 193-196.

17. Gajski, D.D., Jianwen Zhu, Dmer, R., Gerstlauer, A., Shuqing Zhao, *SpecC: Specification Language and Methodology*, Springer, 2000.
18. G. Arnout, "SystemC standard", *Proceedings of the 2000 conference on Asia South Pacific design automation*, ACM Press, 2000, pp. 573-578.
19. S. Vernalde, P. Schaumont, I. Bolsens, "An Object Oriented Programming Approach for Hardware Design", *Proceedings of the IEEE Computer Society Workshop on VLSI'99*, IEEE Computer Society, 1999, pp. 68.
20. K. Ramamritham, K. Arya, "System Software for Embedded Applications", *Proceedings of the 17th International Conference on VLSI Design"*, IEEE Computer Society, 2004, pp. 22.
21. *RC1000-PP Reference Manuals*, Celoxica, 2001.
22. Maxfield, C., *The Design Warrior's Guide to FPGAs: Devices, Tools and Flows*, Elsevier Science, 2004.
23. Söderstrom, T. and Stoica, P., *System Identification*, Prentice-Hall, 1989.
24. Ljung, L., *System Identification*, Prentice-Hall. 1999.
25. J.A. Gómez, M.A. Vega, J.M. Sánchez, "Parametric Identification of Solar Series based on an Adaptive Parallel Methodology", *Journal of Astrophysics and Astronomy* 26, 2005, pp 1-13.
26. Back, T., Fogel, D.B., Michalewicz, Z., *Handbook of Evolutionary Computation*, Oxford University Press, NewYork, 1997.
27. M.A. Vega, J.M. Sánchez, J.A. Gómez, "Advances in FPGA Tools and Techniques", *Microprocessors and Microsystems* 29 (2-3), Elsevier Science, 2005, pp. 47-50.

Part III

Distributed Evolutionary Computation

5

Performance of Distributed GAs on DNA Fragment Assembly

Enrique Alba and Gabriel Luque

[1] Departamento de Lenguajes y Ciencias de la Computación
E.T.S.I. Informática, University of Málaga, 29071, Spain
eat@lcc.uma.es, http://polaris.lcc.uma.es/ eat/
[2] Departamento de Lenguajes y Ciencias de la Computación
E.T.S.I. Informática, University of Málaga, 29071, Spain
gabriel@lcc.uma.es, http://neo.lcc.uma.es/staff/gabriel/index.html

In this work, we present results on analyzing the behavior of a parallel distributed genetic algorithm over different LAN technologies. Our goal is to offer a study on the potential impact in the search mechanics when shifting between LANs. We will address three LANs: a Fast Ethernet network, a Gigabit Ethernet network, and a Myrinet network. We also study the importance of several parameters of the migration policy. The whole analysis will use the DNA fragment assembly problem to show the actual power and utility of the proposed distributed technique.

5.1 Introduction

In practice, optimization problems are often NP-hard, complex, and CPU time-consuming. Although the use of genetic algorithms (GAs) allows to significantly reduce the temporal complexity of the search process (e.g., linear time complexity), times is always a factor for industrial problems. Therefore, parallelism is necessary to reduce not only the resolution time, but also to improve the quality of the provided solutions.

Parallel genetic algorithms differ in the characteristics of their elementary components and in the communication details. In this work, we have chosen a kind of decentralized distributed search because of its popularity and because it can be easily implemented in clusters of machines. In the *distributed GA* (dGA) [6] separate subpopulations evolve independently with sparse exchanges of a given number of individuals with a certain given frequency.

The goal of this work is first study the performance of a parallel GA over three different local area networks: a Fast Ethernet network, a Gigabit

E. Alba and G. Luque: *Performance of Distributed GAs on DNA Fragment Assembly*, Studies in Computational Intelligence (SCI) **22**, 97–115 (2006)
www.springerlink.com

Ethernet network, and a Myrinet network. The two first are very well-known networks. Both have similar characteristics, with the Gigabit Ethernet increasing speed tenfold over Fast Ethernet to 1000 Mbps, or 1 Gbps. Myrinet [8] is a cost-effective and high-performance technology that is also widely used to interconnect clusters of machines in scientific workplaces. Characteristics that distinguish Myrinet from other networks include full-duplex Gigabit/second data rate links, flow control, and low latency. In this study we focus on the influence of several parameter of the migration policy (migration rate and migration gap) in the behavior of a dGA when it is executed over different LANs.

Instead of an academic benchmarking, for performing this study we use a very difficult and appealing problem: the DNA fragment assembly problem. DNA fragment assembly is a technique that attempts to reconstruct the original DNA sequence from a large number of fragments, each one having several hundred base-pairs (bps) long. DNA fragment assembly is needed because current technology, such as gel electrophoresis, cannot directly and accurately sequence DNA molecules longer than 1000 bases. However, most genomes are much longer than this, and new techniques are wanted. For example, a human DNA is about 3.2 billion nucleotides in length and cannot be read at once, so powerful algorithms (hopefully parallel ones) come to scene.

This chapter is organized as follows. First, we present background information about the DNA fragment assembly problem. In the next section, a description of the standard distributed model of GA, in which the population is divided into several islands, is given. In Section 5.4, the details of the implementation are presented. We discuss how to design and implement a distributed genetic algorithm for the DNA fragment assembly problem. In Section 5.5, we test and compare the behavior of the resulting parallel GA over different local area networks, a Fast Ethernet network, a Gigabit Ethernet network, and a Myrinet network. Finally, we summarize our most important conclusions.

5.2 DNA Fragment Assembly Problem

We start this section by giving a vivid analogy to the fragment assembly problem: "Imagine several copies of a book cut by scissors into thousands of pieces, say 10 millions. Each copy is cut in an individual way such that a piece from one copy may overlap a piece from another copy. Assume one million pieces lost and remaining nine million are splashed with ink, try to recover the original text." [17]. We can think of the DNA target sequence as being the original text and the DNA fragments are the pieces cut out from the book. To further understand the problem, we need to know the following basic terminology:

- **Fragment:** A short sequence of DNA with length up to 1000 bps.

- **Shotgun data:** A set of fragments.
- **Prefix:** A substring comprising the first n characters of fragment f.
- **Suffix:** A substring comprising the last n characters of fragment f.
- **Overlap:** Common sequence between the suffix of one fragment and the prefix of another fragment.
- **Layout:** An alignment of collection of fragments based on the overlap order.
- **Contig:** A layout consisting of contiguous overlapping fragments.
- **Consensus:** A sequence derived from the layout by taking the majority vote for each column of the layout.

To measure the quality of a consensus, we can look at the distribution of the coverage. Coverage at a base position is defined as the number of fragments at that position. It is a measure of the redundancy of the fragment data. It denotes the number of fragments, on average, in which a given nucleotide in the target DNA is expected to appear. It is computed as the number of bases read from fragments over the length of the target DNA [18].

$$Coverage = \frac{\sum_{i=1}^{n} length\ of\ the\ fragment\ i}{target\ sequence\ length} \tag{5.1}$$

where n is the number of fragments. TIGR (*The Institute for Genetic Research*) uses the coverage metric to ensure the correctness of the assembly result. The coverage usually ranges from 6 to 10 [13]. The higher the coverage, the fewer the gaps are expected, and the better the result.

5.2.1 DNA Sequencing Process

To determine the function of specific genes, scientists have learned to read the sequence of nucleotides comprising a DNA sequence in a process called DNA sequencing. The fragment assembly starts with breaking the given DNA sequence into small fragments. To do that, multiple exact copies of the original DNA sequence are made. Each copy is then cut into short fragments at random positions. These are the first three steps depicted in Figure 5.1 and they take place in the laboratory. After the fragment set is obtained, traditional assemble approach is followed in this order: overlap, layout, and then consensus. To ensure that enough fragments overlap, the reading of fragments continues until the coverage is satisfied. These steps are the last three steps in Figure 5.1. In what follows, we give a brief description of each of the three phases, namely overlap, layout, and consensus.

Overlap Phase - Finding the overlapping fragments.
This phase consists in finding the best (longest match) between the suffix of one sequence and the prefix of another. In this step, we compare all possible pairs of fragments to determine their similarity. Usually, the dynamic programming algorithm applied to semiglobal alignment is used in this step.

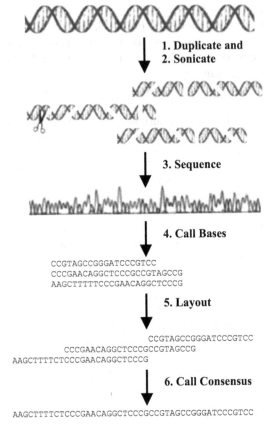

Fig. 5.1. Graphical representation of DNA sequencing and assembly [7]

The intuition behind finding the pairwise overlap is that fragments with a significant overlap score are very likely next to each other in the target sequence.

Layout Phase - Finding the order of fragments based on the computed similarity score. This is the most difficult step because it is hard to tell the true overlap due to the following challenges:

1. Unknown orientation: After the original sequence is cut into many fragments, the orientation is lost. The sequence can be read in either 5' to 3' or 3' to 5'. One does not know which strand should be selected. If one fragment does not have any overlap with another, it is still possible that its reverse complement might have such an overlap.
2. Base call errors: There are three types of base call errors: substitution, insertion, and deletion errors. They occur due to experimental errors in the electrophoresis procedure. Errors affect the detection of fragment overlaps.

Hence, the consensus determination requires multiple alignments in high coverage regions.

3. Incomplete coverage: It happens when the algorithm is not able to assemble a given set of fragments into a single contig.

4. Repeated regions: Repeats are sequences that appear two or more times in the target DNA. Repeated regions have caused problems in many genome-sequencing projects, and none of the current assembly programs can handle them perfectly.

5. Chimeras and contamination: Chimeras arise when two fragments that are not adjacent or overlapping on the target molecule join together into one fragment. Contamination occurs due to the incomplete purification of the fragment from the vector DNA.

After the order is determined, the progressive alignment algorithm is applied to combine all the pairwise alignments obtained in the overlap phase.

Consensus Phase – Deriving the DNA sequence from the layout. The most common technique used in this phase is to apply the majority rule in building the consensus.

Example: We next give an example of this process. Given a set of fragments {F1 = GTCAG, F2 = TCGGA, F3 = ATGTC, F4 = CGGATG}, assume the four fragments are read from 5' to 3' direction. First, we need to determine the overlap of each pair of the fragments by the using semiglobal alignment algorithm. Next, we determine the order of the fragments based on the overlap scores, which are calculated in the overlap phase. Suppose we have the following order: **F2 F4 F3 F1**. Then, the layout and the consensus for this example can be constructed as follows:

```
F2 ->     TCGGA
F4 ->     CGGATG
F3 ->       ATGTC
F1 ->         GTCAG
---------------------
Consensus -> TCGGATGTCAG
```

In this example, the resulting order allows to build a sequence having just one contig. Since finding the exact order takes a huge amount of time, a heuristic such as genetic algorithm can be applied in this step [14, 15, 16].

5.3 Distributed Genetic Algorithms

In the GA field, it is usual to find algorithms implementing panmictic populations, in which selection takes place globally and any individual can potentially mate with any other one in the same single population. The same holds for the replacement operator, where any individual can potentially be removed from the pool and replaced by a new one. In contrast, there exists a different (decentralized) selection/replacement model, in which individuals are arranged spatially, therefore giving place to *structured GAs*. Most other operators, such as recombination or mutation, can be readily applied to these two models.

There exists a long tradition in using structured populations in EC (Evolutionary Computation), especially associated to parallel implementations [9]. Among the most widely known types of structured GAs, the *distributed* (dGA) and *cellular* (cGA) ones are popular optimization procedures [6].

Decentralizing a single population can be achieved by partitioning it into several subpopulations, where island GAs are run performing sparse exchanges (migrations) of individuals (dGAs), or in the form of neighborhoods (cGAs). These two GA types, along with a panmictic GA, are schematically depicted in Figure 5.2.

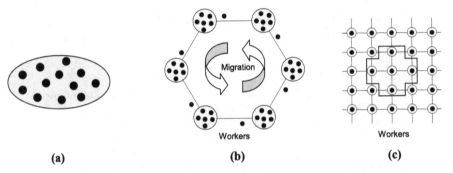

(a)   **(b)**   **(c)**

Fig. 5.2. A panmictic GA has all its individual –black points– in the same population and, thus, (a) each of them can potentially mate with any other. Structuring the population usually leads to a distinction between (b) dGAs and (c) cGAs

In this chapter, we focus our study on dGAs. In distributed GAs, the population is structured into smaller subpopulations relatively isolated one from the others. Parallel GAs based on this paradigm are sometimes called multipopulation or multi-deme GAs. Besides its name, the key characteristic of this kind of algorithm is that individuals within a particular subpopulation (or island) can occasionally migrate to another one. This paradigm is illustrated in Figure 5.2 (b). Note that the communication channels shown are notional; specific assignments are made as a part of the GA's migration strategy and are mapped to some physical network.

It has to be mentioned that structure GAs are new kind of GAs [2] and that any of these models admit several different implementations on parallel computers (model vs. implementation).

Conceptually, the overall GA population is partitioned into a number of independent, separate subpopulations (or demes). An alternative view is that of several small, separate GAs executing simultaneously. Regardless, individuals occasionally migrate between one particular island and its neighbors, although these islands usually evolve in isolation for the majority of the GA run. Here, genetic operators (selection, mutation, and recombination) take place within each island, which means that each island can search in very different regions of the whole search space with respect to the others. Each island could also have different parameter values (heterogeneous GAs [4]). The distributed model requires the identification of a suitable migration policy. The main parameters of the migration policy include the following ones:

- *Migration Gap.* Since a dGA usually makes sparse exchanges of individuals among the subpopulations, we must define the migration gap which is the number of steps in every subpopulation between two successive exchanges (steps of isolated evolution). Migration can occur in every subpopulation either periodically or governed by a given probability P_M.
- *Migration Rate.* This parameter determines the number of individuals that undergo migration in every exchange. Its value can be given as a percentage of the population size or else as an absolute value.
- *Selection/Replacement of Migrants.* This parameter defines how to select emigrant solutions, and which solutions have to be replaced by the in-migrants. It is very common in parallel distributed GAs to use the same selection/replacement operators for dealing with migrants.
- *Topology.* This parameter defines the neighbor of each island, i.e., the target islands for the exchanges of information. The traditional nomenclature divides parallel GAs into *stepping-stone* and *island* models, depending on whether individuals can freely migrate to any subpopulation or if they are restricted to migrate to geographically nearby islands, respectively.

5.4 DNA Fragment Assembly Using a Distributed GA

In this section we show how dGAs can be used to solve the DNA fragment assembly problem.

Solution Representation

We use the permutation representation with integer number encoding. A permutation of integers represents a sequence of fragment numbers, where successive fragments overlap. The solution in this representation requires a list of fragments assigned with a unique integer ID. For example, 8 fragments will need eight identifiers: 0, 1, 2, 3, 4, 5, 6, 7. The permutation representation requires special operators to make sure that we always get legal (feasible) solutions. In order to maintain a legal solution, the two conditions that must be satisfied are (1) all fragments must be presented in the ordering, and (2) no duplicate fragments are allowed in the ordering. For example, one possible ordering for 4 fragments is 3 0 2 1. It means that fragment 3 is at the first position and fragment 0 is at the second position, and so on.

Fitness Function

In the DNA fragment assembly problem, the fitness function measures the multiple sequences alignment quality and finds the best scoring alignment. Parsons, Forrest, and Burks mentioned two different fitness functions [15]. In this chapter we use the first proposed one. This fitness function sums the overlap score for adjacent fragments in a given solution. When this fitness function is used, the objective is to maximize such a score. It means that the best individual will have the highest score.

$$F(l) = \sum_{i=0}^{n-2} w(f[i]f[i+1]) \tag{5.2}$$

The overlap score in F is computed using the semiglobal alignment algorithm.

Recombination Operator

For our experimental runs, we use the order-based crossover (OX). This operator was specifically designed for tackling problems with permutation representations. The order-based crossover operator first copies the fragment ID between two random positions in Parent1 into the offspring's corresponding positions. It then copies the rest of the fragments from Parent2 into the offspring in the relative order presented in Parent2. If the fragment ID is already present in the offspring, then it skips that fragment. The method preserves the feasibility of every string in the population.

Mutation Operator

For our experimental runs we use the swap mutation operator. This operator randomly selects two positions from a permutation and then swaps the two resulting fragments. Since this operator does not introduce any duplicate value in the permutation, the offspring is always feasible. Swap mutation operator is suitable for permutation problems like ordering fragments.

Selection Operator

In this work, we use ranking selection mechanism [19], in which the GA first sorts the individuals based on the fitness and then selects the individuals with the best fitness score until the specified population size is reached. Note that the population size will grow whenever a new offspring is produced by crossover or mutation. The use of ranking selection is here preferred over other selections such as fitness proportional selection for statistical reasons [12].

Before moving on to Section 5.5 in which we analyze the effects of several configurations of migration rates and gaps on different networks, we introduce MALLBA, the software library which we are using in this work.

5.4.1 The MALLBA Project

The MALLBA research project [5] is aimed at developing a library of algorithms for optimization that can deal with parallelism in a user-friendly and, at the same time, efficient manner. Its three target environments are sequential, LAN and WAN computer platforms. All the algorithms described in the next section are implemented as *software skeletons* (similar to the concept of software patterns) with a common internal and public interface. This permits fast prototyping and transparent access to parallel platforms.

MALLBA skeletons distinguish between the concrete problem to be solved and the solver technique. Skeletons are generic templates to be instantiated by the user with the features of the problem. All the knowledge related to the solver method (e.g., parallel considerations) and its interactions with the problem are implemented by the skeleton and offered to the user. Skeletons are implemented by a set of *required* and *provided* C++ classes that represent an abstraction of the entities participating in the solver method:

- **Provided Classes**: They implement internal aspects of the skeleton in a problem-independent way. The most important *provided* classes are `Solver` (the algorithm) and `SetUpParams` (setup parameters).
- **Required Classes**: They specify information related to the problem. Each skeleton includes the `Problem` and `Solution` required classes that encapsulate the problem-dependent entities needed by the solver method. Depending on the skeleton other classes may be required.

Therefore, the user of a MALLBA skeleton only needs to implement the particular features related to the problem. This speeds considerably the creation of new algorithms with minimum effort, especially if they are built up as combinations of existing skeletons (*hybrids*).

The parallel skeletons of MALLBA use `NetStream` [3] to perform the communication phase. `NetStream` is a light middleware layer over the standard MPI [11]. The MALLBA library is publicly available at `http://neo.lcc.uma.es/mallba/easy-mallba/`.

By using this library, we were able to perform a quick coding of algorithmic prototypes to cope with the inherent difficulties of the DNA fragment assembly problem.

5.5 Experimental Results

In this section we analyze the behavior of a distributed GA when it is executed over several local area networks: Fast Ethernet, Gigabit Ethernet, and Myrinet networks. In this aim we begin by performing an experimental set of tests for several migration gaps and rates. First, we describe the parameters used in these experiments, and later we analyze the results.

A target sequence with accession number BX842596 (GI 38524243) was used as the problem instance in this work. It was obtained from the NCBI web site (http://www.ncbi.nlm.nih.gov). It is the sequence of a Neurospora crassa (common bread mold) BAC, and is 77,292 base pairs long. To test and analyze the performance of our algorithm, we generated a problem instance with GenFrag [10]. The problem instance, 842596_4, contains 442 fragments with average fragment length of 708 bps and coverage 4.

In the experiments we use different values of migration gap (1, 2, 16, 128, 512 and 1024 generations) and rate (1, 2, 16, 32 and 64 individuals). Notice that a low gap value (e.g., 1) means high coupling, while a high value (e.g., 64) means loose coupling. The migration in dGAs occurs asynchronously in a unidirectional ring manner, sending randomly chosen individuals to the neighbor sub-population. The target population incorporates these individuals only if they are better than its presently worst solution. In the first phase, we use a $(\mu + \mu)$-dGA with 8 islands (an island per processor). In the latter phase, we change the number of islands/processors and we analyze a dGA with 4 and 2 islands. The remainder parameters are shown in the Table 5.1. In these experiments the hardware and software environments where the algorithms are executed are influent issues in parallel optimization. These parameters are shown in the Table 5.2. Because of the stochastic nature of the dGA, we perform 30 independent runs of each test to gather statistical meaningful experimental data.

Table 5.1. Parameter settings

Parameter	Value
Subpopulation size	1024/Number of processors
Representation	Permutation (442 integers)
Crossover operator	Order-based ($P_C = 0.8$)
Mutation operator	Swap ($P_M = 0.2$)
Selection method	Ranking selection
Migration Rates	1, 2, 16, 32, and 64
Migration Gaps	1, 2, 64, 128, 512, and 1024

Table 5.2. Hardware and software environments

Parameter	Value
Processor	PIV 2.4 GHz
Main Memory	256 MB
Linux Kernel	2.4.19-6 (SuSE)
g++ version	3.2
MPICH version	1.2.2.3

Tables 5.3, 5.4, and 5.5 show the mean execution times of the dGA (having 8 islands) over a Fast Ethernet network, a Gigabit Ethernet network, and a Myrinet network, respectively. For clarity, we also plot several of these values in Figure 5.3.

Table 5.3. Mean execution time (μsec) of a dGA having 8 processors over a Fast Ethernet network

				gap			
		1	2	64	128	512	1024
	1	37269683	34967250	33963737	33799607	33890313	33836158
	2	37092247	35766270	34101287	33730040	33998213	33843180
rate	16	58294070	47610630	35971320	33784687	33962590	33860780
	32	89803950	62581517	37543203	34065667	33904353	33915620
	64	175254844	98745559	40857867	34482773	33872323	33956670

Table 5.4. Mean execution time (μsec) of a dGA having 8 processors over a Gigabit Ethernet network

				gap			
		1	2	64	128	512	1024
	1	35842227	34923417	33772577	33681037	33676473	33826972
	2	36481183	35312083	33746657	33668543	33598087	33647733
rate	16	51429240	44642810	33895067	33769983	33723950	33726360
	32	64471883	50845050	34250270	33989930	33709327	33690097
	64	80958070	59404767	34678460	34160270	33732820	33710255

Table 5.5. Mean execution time (μsec) of a dGA having 8 processors over a Myrinet network

				gap			
		1	2	64	128	512	1024
	1	35888297	34778717	33666510	33765507	33720580	33716997
	2	36137960	35003820	33644037	33579163	33656493	33647457
rate	16	50569180	45055760	33979783	33874037	33735097	33685963
	32	65492757	51465307	34434347	33982707	33754890	33694487
	64	83340960	60676630	34518957	34027937	33801083	33778647

Let us now discuss the results found in these tables. First, we can notice that, for small values of migration gap, the execution time is increased with respect to high gaps. As expected, when we send messages very frequently (information exchanges occur each one or two generations), the communication overhead is high and then, the total execution time also is high. Second, we

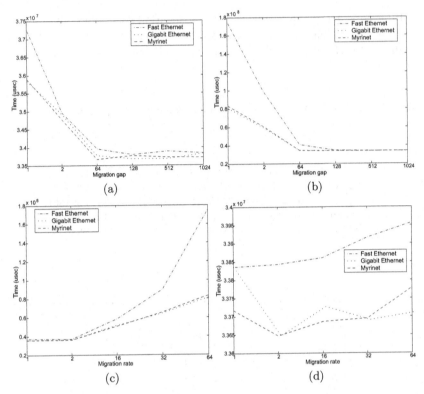

Fig. 5.3. Graphical representation of the mean execution (μsec) of a dGA with (a) migration rate = 1, (b) migration rate = 64, (c) migration gap = 1, and (d) migration gap = 1024

can observe that the larger the migration rate, the highest the runtime execution. This is not surprising, since we are exchanging data packages very large with several individuals in each message. The average size of a data packet is: 1.7 KB, 3.4 KB, 27.6 KB, 55.2 KB, and 110 KB (for migration rates = 1, 2, 16, 32, and 64, respectively).

Now, we analyze the performance of the distributed algorithm over all the networks. We can observe that the Fast Ethernet network is the slowest in all the cases, although for the configurations with low communications (small rate and high gaps), its execution time is similar to the other networks (for example, see right values of Figures 5.3(a) and 5.3(b)). There are not significant differences between the Gigabit Ethernet times and the Myrinet ones. In general, Myrinet network is better than Gigabit Ethernet one for configurations of the algorithm where there is sparse communications between the islands (values of migration rate smaller than 16 individuals and values of migration gap greater than 64 generations). However, the more intense the communication phase is, the more similar Myrinet and Gigabit Ethernet

Table 5.6. Ratio between the communication time and the total runtime for a Fast Ethernet network

		gap					
		1	2	64	128	512	1024
	1	2.2673	1.5839	0.3195	0.14822	0.12907	0.11877
	2	3.3623	1.9471	0.3861	0.15295	0.1273	0.12076
rate	16	32.2	21.951	4.9972	0.39005	0.25331	0.15037
	32	46.921	32.135	6.7842	1.0607	0.30336	0.21534
	64	62.818	48.974	12.015	1.8709	0.39075	0.30922

networks are. In the limit (i.e., high migration rates and low migration gaps) we have found that Gigabit Ethernet executions are even more efficient than Myrinet ones. We think that the reason for this behavior is that the OS or the communication software layer (MPI) do not allow to use completely all the features of this network.

Table 5.7. Ratio between the communication time and the total runtime for a Gigabit Ethernet network

		gap					
		1	2	64	128	512	1024
	1	2.1843	1.3771	0.23906	0.14338	0.12619	0.11838
	2	3.1538	1.7501	0.25413	0.15678	0.13078	0.12024
rate	16	29.014	20.814	0.55587	0.31725	0.1659	0.13569
	32	42.069	28.958	1.7742	0.95777	0.31941	0.20433
	64	55.247	39.093	2.4848	1.2916	0.39944	0.2393

Table 5.8. Ratio between the communication time and the total runtime for a Myrinet network

		gap					
		1	2	64	128	512	1024
	1	2.7112	1.4757	0.18691	0.1439	0.12697	0.1175
	2	3.5726	1.9975	0.21191	0.16187	0.12788	0.11827
rate	16	27.75	20.862	1.2862	0.73792	0.18074	0.16755
	32	41.047	28.37	1.6797	0.91508	0.34357	0.19961
	64	56.626	40.119	2.3893	1.2745	0.52415	0.2338

Tables 5.6, 5.7, and 5.8 show the proportion between the time that the algorithm is in a communication phase and the total runtime for all the networks. Myrinet and Gigabit Ethernet networks obtain a very similar value, and the Fast Ethernet gets always a slightly larger percentage. This means that the Fast Ethernet loses more time in the exchanging data packets than

the other networks. In all networks, for low values of migration rate (lower or equal than two) and high values of migration gap (higher or equal than 128), the communication time is negligible (it represents less than 3% of the complete execution time) but if we increase the rate or decrease the gap, the communication phase dominates the computational effort. Even, in some configuration, the algorithm spends more time in communication phase (55-62% of the total execution time) than in all the other phases summed together.

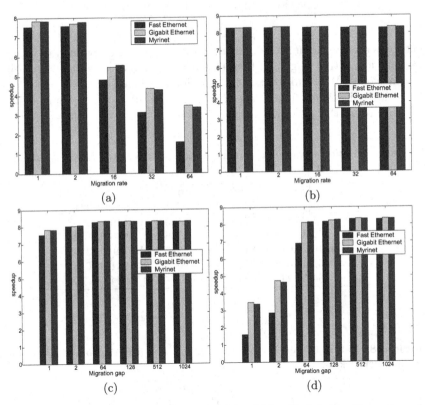

Fig. 5.4. Weak (versus panmixia) speedup of a dGA with (a) migration rate = 1, (b) migration rate = 64, (c) migration gap = 1, and (d) migration gap = 1024

Let us now turn to a deeper analysis of times. For this, the most important measure of a parallel algorithm is the *speedup*. This metric compares two times: ratio between sequential and parallel times. The main difficulty with that measure is that researchers do not agree on the meaning of *sequential time*. For example, in the Figure 5.4 we compare the parallel algorithm (eight processors) against the canonical sequential version (weak versus panmixia speedup [1]). We can notice that in a dGA with sparse communications the speedup is very good, even it is super-linear in many cases (see Figures 5.4(b) and 5.4(c)). However, the speedup decreases quickly when we use high values of migration rate and low values of migration gap. For configurations with low communications, all networks obtain a similar and very good speedup, but for the remainder configurations the algorithm have a considerable loss of efficiency (see right values of Figure 5.4(a) and left ones of Figure 5.4(d)). This loss is more important in the Fast Ethernet executions than in the Myrinet and Gigabit ones.

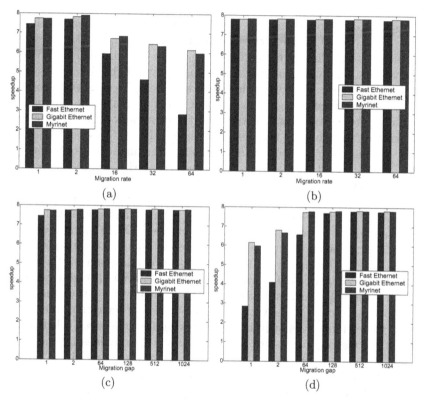

Fig. 5.5. Weak (orthodox) speedup of a dGA with (a) migration rate = 1, (b) migration rate = 64, (c) migration gap = 1, and (d) migration gap = 1024

The previous comparison can be further improved, because the two algorithms are different. In the Figure 5.5, we perform a more fair comparison; this time we will compare the runtime of the parallel algorithm running on one processor against the runtime of the same algorithm running on 8 processors (orthodox speedup [1]). Anyway, we obtain similar conclusions: configurations with low communication obtain nearly linear speedup (before it was super-linear) and configurations with high communication obtain a moderate speedup but still better than in the preceding no orthodox results (Figure 5.4).

Again, this loss of efficiency is most clear in the Ethernet results than in the other networks.

In the preceding analysis we dealt with a distributed genetic algorithm having eight islands (eight processors). Now, we will extend the study to cover other number of islands and processors. Our aim is to check whether the previously observed behavior with eight processors still holds when shifting to a dGA having two o four islands/processors. For clarity, we simplify this study and only analyze bounding values of migration ratio (1 and 64) and migration gap (1 and 1024).

As we did before, we analyze the execution time, the communication cost and the (orthodox) speedup over all configurations and networks. These value are shown in the Figure 5.6. In all these figures, we can distinguish two behaviors: the first one, when the communication phase is light (left values of each graph), the differences between the networks is negligible, and the second one, for configurations with high comunication, the diffences are more important, specially between the Fast Ethernet and the other networks. As it occurs in the previous experiments, the time differences (Figures 5.6(a) and 5.6(b)) between the networks are only significant when the algorithm exchanges data very frequently and the size of the messages is large (i.e., with tight communication patterns). However these differences between the networks are smaller when we decrease the number of islands. This occurs since a smaller number of islands means a smaller number of messages in the network, and there are a lower number of collisions in the Ethernet, hence allowing to improve performance. The behavior of the speedup (Figures 5.6(e) and 5.6(f)) is also similar to preceding results, i.e., when the communication phase is light, the speedup is nearly linear, and it deteriorates for configurations with high communication. A noticeable result is that the sequential time is lower than the parallel time for one configuration (with the highest value of migration rate and the lowest value of migration gap) of a dGA having two processors, what remains us that parallel algorithms are not always better than sequential ones.

Let us now summarize the results. Here, we have checked the behavior of a distributed GA over three LAN technologies using different parameter settings. We have observed that the performance of the dGA over different networks is very dependant of the values of the migration rate, the migration frequency, and the number of processors. We have noticed when the communication phase is light, i.e., the value of the migration rate is low and/or the value of the migration gap is high, or when we use a small number of processors, the performance is quite similar in all the networks. For the remainder configurations (i.e., high communications ones), the Fast Ethernet performs worse than Myrinet and Gigabit. Myrinet performs slightly better than Gigabit in configurations with low communication, while Gigabit is slightly better in the other cases. However we cannot appreciate important differences between Myrinet and Gigabit in any configurations.

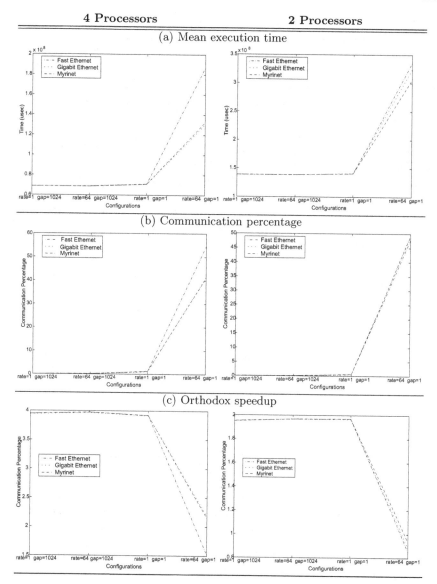

Fig. 5.6. Summary of the results of a dGA having two and four processors

5.6 Summary

In this chapter we analyze the behavior of a parallel distributed genetic algorithm over different networks and study the effects of several parameters of the migration policy in the execution time.

We have shown that differences between networks vanish when the algorithm performs sparse communications, i.e., it does not send message very frequently (each 16 or more generations) and the size of these messages is medium-small (packages smaller or equal than 3.4 KB). In highly coupled dGAs, we have observed that the Fast Ethernet network provokes a clearly higher overhead than Myrinet and Gigabit. However, we have also noticed that these differences are reduced when we use a small number of processors, due to a smaller saturation of the network.

In general, we have also observed that the executions times using Myrinet and Gigabit are similar. So we need still to test the algorithms to find out in which cases Myrinet performs better to justify its expensive cost in the market, since its low latency does not shown up as impact factor in this study.

As a future work we plan to extend this paper to account for the effects of other parameters such as the migration topology. Also, we want to check the performance of the networks when using other parallel models of genetic algorithms different from the distributed one such as the master-slave model or the cellular model, where the communication phase is more expensive.

Acknowledgments

The authors are partially supported by the Ministry of Science and Technology and FEDER under contract TIC2002-04498-C05-02 (the TRACER project).

References

1. E. Alba. Parallel evolutionary algorithms can achieve super-linear performace. *Information Processing Letters*, 82:7–13, 2002.
2. E. Alba. *Parallel Metaheuristics: A New Class of Algorithms*. Wiley, 2005.
3. E. Alba, C. Cotta, M. Díaz, E. Soler, and J.M. Troya. MALLBA: Middleware for a geographically distributed optimization system. Technical report, Dpto. Lenguajes y Ciencias de la Computación, Universidad de Málaga (internal report), 2000.
4. E. Alba, F. Luna, and A.J. Nebro. Advances in parallel heterogeneous genetic algorithms for continuous optimization. *International Journal of Applied Mathematics and Computer Science*, 14(3):317–333, 2004.
5. E. Alba and the MALLBA Group. MALLBA: A library of skeletons for combinatorial optimisation. In R. Feldmann B. Monien, editor, *Proceedings of the Euro-Par*, volume 2400 of *LNCS*, pages 927–932, Paderborn (GE), 2002. Springer-Verlag.
6. E. Alba and M. Tomassini. Parallelism and Evolutionary Algorithms. *IEEE Transactions on Evolutionary Computation*, 6(5):443–462, October 2002.
7. C.F. Allex. *Computational Methods for Fast and Accurate DNA Fragment Assembly*. UW technical report CS-TR-99-1406, Department of Computer Sciences, University of Wisconsin-Madison, 1999.

8. N.J. Boden, D. Cohen, R.E. Felderman, A.E. Kulawik, C.L. Seitz, J. Seizovic, and W. Su. Myrinet A Gigabit Per Second Local Area Network. *IEEE Micro*, pages 29–36, 1995.

9. E. Cantú-Paz. *Efficient and Accurate Parallel Genetic Algorithms*. Kluwer Academic Press, 2000.

10. M.L. Engle and C. Burks. Artificially generated data sets for testing DNA fragment assembly algorithms. *Genomics*, 16, 1993.

11. Message Passing Interface Forum. MPI: A message-passing interface standard. Technical Report UT-CS-94-230, 1994.

12. D.E. Goldberg and K. Deb. A comparative analysis of selection schemes used in genetic algorithms. In G.J.E. Rawlins, editor, *Foundations of Genetic Algorithms*, pages 69–93. Morgan Kaufmann, 1991.

13. S. Kim. *A structured Pattern Matching Approach to Shotgun Sequence Assembly*. PhD thesis, Computer Science Department, The University of Iowa, 1997.

14. C. Notredame and D.G. Higgins. SAGA: sequence alignment by genetic algorithm. *Nucleic Acids Research*, 24:1515–1524, 1996.

15. R. Parsons, S. Forrest, and C. Burks. Genetic algorithms, operators, and DNA fragment assembly. *Machine Learning*, 21:11–33, 1995.

16. R. Parsons and M.E. Johnson. A case study in experimental design applied to genetic algorithms with applications to DNA sequence assembly. *American Journal of Mathematical and Management Sciences*, 17:369–396, 1995.

17. P.A. Pevzner. *Computational molecular biology: An algorithmic approach*. The MIT Press, London, 2000.

18. J. Setubal and J. Meidanis. *Introduction to Computational Molecular Biology*, chapter 4 - Fragment Assembly of DNA, pages 105–139. University of Campinas, Brazil, 1997.

19. D. Whitely. The GENITOR algorithm and selection pressure: Why rank-based allocation of reproductive trials is best. In J.D. Schaffer, editor, *Proceedings of the Third International Conference on Genetic Algorithms*, pages 116–121. Morgan Kaufmann, 1989.

6

On Parallel Evolutionary Algorithms
on the Computational Grid

N. Melab, E-G. Talbi and S. Cahon

Laboratoire d'Informatique Fondamentale de Lille
UMR CNRS 8022, INRIA Futurs - DOLPHIN Project
Cité scientifique - 59655, Villeneuve d'Ascq cedex - France
{melab,talbi,cahon}@lifl.fr

In this chapter, we analyze the major traditional parallel models of evolutionary algorithms. The analysis is a contribution in parallel evolutionary computation as unlike previously published studies it is placed within the context of grid computing[1]. The objective is to visit again the parallel models in order to allow their adaptation to grids taking into account the characteristics of such execution environments in terms of volatility, heterogeneity, large scale and multi-administrative domain. The proposed study is a part of a methodological approach for the development of frameworks dedicated to the reusable design of parallel EAs transparently deployable on computational grids. We give an overview of such frameworks and present a case study related to ParadisEO-CMW which is a porting of ParadisEO onto Condor and MW allowing a transparent deployment of the parallel EAs on grids.

6.1 Introduction

Near-optimal algorithms such as Evolutionary Algorithms (EAs) allow to significantly reduce the time and space complexity in solving combinatorial optimization problems. However, they are unsatisfactory to tackle large problems without the resort to parallel computing on computational grids which has recently been revealed as a powerful way to deal with time-intensive problems [1]. A computational grid is a scalable pool of heterogeneous and dynamic resources geographically distributed across multiple administrative domains

---

[1] This work is part of the *CHallenge in Combinatorial Optimization (CHOC)* project supported by the *National French Research Agency (ANR)* through the *Hign-Performance Computing and Computational Grids (CIGC)* programme

N. Melab et al.: *On Parallel Evolutionary Algorithms on the Computational Grid*, Studies in Computational Intelligence (SCI) **22**, 117–132 (2006)
www.springerlink.com

and owned by different organizations. The characteristics of such environment may have a great impact on the design of grid-based EAs.

Three major parallel models of EAs are described in this chapter: the *parallel island model*, the *parallel evaluation of a population* and the *parallel evaluation of a single solution* [2]. Briefly, they allow respectively the simultaneous deployment of cooperative EAs and the parallel evaluation of the population of an EA and each of its solutions in an independent and farmer-worker way. These traditional parallel models are re-visited and analyzed taking into account the characteristics of computational grids.

The design of grid-aware EAs often involves the cost of a sometimes painful apprenticeship of parallelization techniques and grid computing technologies. In order to free from such burden the programmers who are unfamiliar with those advanced features, frameworks must provide the implementation of the three models and allow their transparent exploitation and deployment on computational grids. In this chapter, we give an overview of the frameworks dedicated to the design of parallel EAs in general and grid-based ones in particular. The ParadisEO-CMW framework [3] which is a porting of ParadisEO [2] onto Condor [4] and MW [5] is presented as a case study of such frameworks. ParadisEO-CMW allows the design of parallel EAs with a transparent deployment of the three models on computational grids.

The organization of the remainder of the chapter is the following. Section 2 highlights the principles of EAs. In Section 3, we analyze the major associated parallel models within the context of computational grids. Section 4 presents a brief overview of frameworks for the design of parallel EAs particularly on grids. A case study related to the ParadisEO-CMW is also provided. In Section 5, a conclusion of the chapter is drawn.

6.2 Principles of Evolutionary Algorithms

Evolutionary Algorithms [6] (see Algorithm 6.1) are population-based meta-heuristics based on the iterative application of stochastic operators on a population of candidate solutions. At each iteration, individuals are selected from the population, paired and recombined in order to generate new ones which replace other individuals selected from the population either randomly or according to a selection strategy.

The components of an EA are mainly: the way the population is initialized, the selection strategy, the replacement approach, and the continuation/stopping criterion. The initial population is either generated at random or provided by an heuristic. The selection strategy aims at fostering better solutions, and can be either *blind* (*stochastic*), meaning individuals are selected randomly, or *intelligent* (*deterministic*) such is tournament, wheel, etc. The replacement approach allows to withdraw individuals selected according to a given selection strategy. The continuation/stopping criterion is evaluated at the end of each generation and the evolution process is stopped if it is satisfied.

Algorithm 6.1 EA pseudo-code

Generate($P(0)$);
$t := 0$;
while not Termination_Criterion($P(t)$) **do**
    Evaluate($P(t)$);
    $P'(t)$:= Selection($P(t)$);
    $P'(t)$:= Apply_Reproduction_Ops($P'(t)$);
    $P(t+1)$:= Replace($P(t)$, $P'(t)$);
    $t := t + 1$;
endwhile

In the Pareto-oriented multi-objective context, the provided result is not a single solution but a set of *non-dominated* solutions called *Pareto Front*. A solution is said non-dominated if it is at least equal to other solutions on all the objectives being optimized, and there exists at least one objective for which it is strictly better. The structure of the evolutionary algorithm remains the same as presented above but some adaptations are required mainly for the evaluation and selection steps. The evaluation phase includes in addition to the computation of a fitness vector (of values associated with the different objectives) the calculation of a global value based on this latter. A new function is thus required to transform the fitness vector into a scalar value defining the quality of the associated individual. Such scalar value is used in other parts of the algorithm particularly the selection phase (*ranking*).

The selection process is often based on two major mechanisms: *elitism* and *sharing*. They allow respectively the convergence of the evolution process to the best Pareto front and to maintain some diversity of the potential solutions. The elitism mechanism makes use of a second population called a *Pareto archive* that stores the different non-dominated solutions generated through the generations. Such archive is updated at each generation and used by the selection process. Indeed, the individuals on which the variation operators are applied are selected either from the Pareto archive, from the population or from both of them at the same time. The sharing operator maintains the diversity on the basis of the similarity degree of each individual compared to the others. The similarity is often defined as the euclidean distance in the objective space.

6.3 Parallel EAs on the Computational Grid

The reproductive cycle of an EA on long individuals and/or large populations requires a large amount of computational resources especially for real-world problems. In general, evaluating a fitness function for every individual is frequently the most costly operation of the EA. Consequently, a variety of algorithmic issues are being studied to design efficient EAs. These issues usually consist of defining new operators, hybrid algorithms, parallel models, and so on. Parallelism on computational grids is particularly inevitable when the problems at hand are large size ones.

The proliferation of research and industrial projects on grid computing is leading to the proposition of several sometimes confusing definitions of the grid concept [7, 8]. As a consequence, some articles such as [9, 10] are especially dedicated to the analysis of these definitions. It is thus important to clearly define the context on which a work is focused when it is related to grid computing. In this chapter, the targeted architecture is the computational grid as defined in [8]. A computational grid is a scalable pool of heterogeneous and dynamic resources geographically distributed across multiple administrative domains and owned by different organizations. The characteristics could be summarized as follows:

- *The grid includes multiple autonomous administrative domains*: The users and providers of resources are clearly identified. This allows to reduce the complexity of the security issue, however, the firewall traversal remains a critical problem to deal with. In global computing middlewares based on the large scale cycle stealing such as XtremWeb [12], the problem is solved in a natural way as communications are initiated from "inside the domains".
- *The grid is heterogeneous*: The heterogeneity in a grid is intensified by its large number of resources belonging to different administrative domains. The emergence of data exchange standards and Java-based technologies such as RMI allows to deal with the heterogeneity issue.
- *The grid has a large scale*: the grid has a large number of resources growing from hundreds of integrated resources to millions of PCs. The design of performant and scalable grid applications has to take into account the communication delays.
- *The grid is dynamic*: The dynamic temporal and spatial availability of resources is not an exception but a rule in a grid. Due to the large scale nature of the grid the probability of some resource failing is high. Such characteristic highlights some issues such as dynamic resource discovery, fault tolerance, and so on.

The gridification of parallel EAs requires to take into account at the same time the characteristics and underlined issues of the computational grids and the parallel models. Some of the issues related to grids may be solved by middlewares [5, 11, 12] allowing to hide their inherent complexity to the users. The number of issues that could be solved in a transparent way for the users depends on the middleware at hand. The choice of this later is crucial for performance and ease of use.

Three major parallel models for EAs stand out in the literature: the *parallel cooperative island model*, the *parallel evaluation of the population model*, and the *parallel evaluation of a single solution model*. To the best of our knowledge, these models have never been studied within the context of computational grids. Indeed, the studies proposed in the literature do not take into account the characteristics of computational grids presented below.

6.3.1 The Island Model

The island model (Figure 6.1) consists in running simultaneously several EAs that exchange genetic stuff to compute better and robust solutions. The objective of the model is to delay the global convergence, especially when the EAs are heterogeneous regarding the variation operators or the machines they are being executed. The EAs of the island model are either *homogeneous* or *heterogeneous* depending on whether their parameters such as the variation operators are identical or not. In the following, we will use in an interchangeable way the terms island, its associated EA and population, and the machine that hosts them.

The exchange of genetic stuff between EAs is called *migration process*, and can be either *synchronous* or *asynchronous*. Migration occurs at the end of each generation following the replacement phase. The migration process is based on a policy defined by the following parameters: the migration decision criterion, the exchange topology, the number of emigrants, the emigrants selection policy, and the replacement/integration policy. In the multi-objective context, the individuals are selected either from the population, the Pareto archive or from the combination of the two. The design of the model is generic and could be completely provided inside a software platform for the design of parallel EAs. The user will indicate if necessary the values of the different parameters of the model.

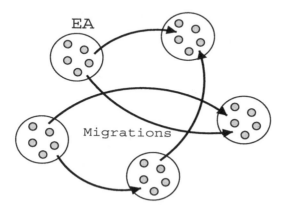

Fig. 6.1. The parallel island model

- **Migration decision criterion**: The migration of individuals between islands can be decided either in a *blind* (periodic or probabilistic) way or according to an *intelligent* criterion. Periodic migration occurs in each island after a fixed number of generations (migration frequency). Probabilistic migration consists in performing a migration operation after each generation with a user-defined probability. Conversely, intelligent migrations are guided by the quality improvement of the population. An improvement threshold must be provided, and if the improvement between two successive generations is below that threshold a migration operation is triggered. The specification of the different parameters associated with the blind or intelligent migration decision criterion (migration frequency/probability and improvement threshold) is particularly crucial on a computational grid. Indeed, due to the heterogeneous nature of this latter these parameters must be specified for each island in accordance with the machine it is hosted on.

- **Migration topology**: The migration topology indicates for each island its neighbor(s) regarding the migration of individuals i.e. the source/destination island(s) of its emigrants/immigrants. Several research works [13, 14, 15] has been dedicated to the study of the impact of the topology on the quality of the provided results, and show that cyclic graphs are better. The ring and hypercube topologies are particularly often used. On a volatile computational grid, it is difficult to efficiently maintain such topologies. Indeed, the disappearance of a given island requires a dynamic reconfiguration of the topology. Such reconfiguration is costly and makes the migration process inefficient. In [15, 16], a cooperation between a set of EAs has been experimented without any topology i.e. the target island of emigrants is selected at random. The experimental results show that such topology allows a significant improvement of the robustness and quality of solutions. The random topology is therefore thinkable and even commendable in a computational grid context.

- **Number of emigrants**: The number of emigrants can be expressed as a fixed or variable number of individuals, or as a percentage of individuals from the population or the Pareto archive (in the multi-objective context). The choice of the value of such parameter is crucial. Indeed, if it is low the migration process will be less efficient as the islands will have the tendency to evolve in an independent way. Conversely, if the number of emigrants is high the communication cost will be exorbitant particularly on a computational grid, and the EAs will likely to converge to the same solutions.

- **Selection policy of emigrants**: The selection policy of emigrants indicates for each island in an *elitist* or *stochastic* way the individuals to be migrated. The stochastic or random policy does not guarantee that the best individuals will be selected, but its associated computation cost is lower. The elitist strategy (wheel, rank, tournament or uniform sampling) allows the selection of the best individuals. In the multi-objective context, the individuals are selected from either the population, the Pareto archive or both of them. The characteristics of computational grids have any influence on the the selection policy of emigrants.

- **Replacement/Integration policy of immigrants**: Symmetrically to the selection policy, the replacement/integration policy of immigrants indicates in a random or elitist way the local individuals to be replaced by the newcomers. In the multi-objective island model, different strategies are possible. For instance, the newcomers replace individuals selected randomly from the local population that do not belong to the local Pareto archive. Another strategy consists in ranking and grouping the individuals of the local population into Pareto fronts using the non-dominance relation. The solutions of the worst Pareto front are thus replaced by the new arrivals. One can also make use of the technique that consists in merging the immigrant Pareto front with the local one, and the result constitutes the new local Pareto archive.

The implementation of the island model is either *asynchronous* or *synchronous*. The asynchronous mode associates with each EA an immigration decision criterion which is evaluated at the end of each iteration of the EA from the state of its population or its Pareto archive. If the criterion is satisfied the neighbors of the EA are requested to send their emigrants. The migration requests are managed by the destination EAs within an undetermined delay. The reception and integration of newcomers is thus performed during the next iterations. However, in a computational grid context, due the material and/or software heterogeneity issue the EAs of the island model could be at different evolution stages leading to the *non-effect* and/or *super-individual* problem. Indeed, the arrival of poor solutions in a population or Pareto archive at a very advanced stage will not bring any contribution as these solutions will likely not be integrated. In the opposite situation, the insular cooperation will lead to a premature convergence. From another point of view, as it is non-blocking the model is more performant and fault tolerant to such a degree a threshold of wasted migrations is not exceeded.

In the synchronous mode, the EAs perform a synchronization operation at the end of each iteration by exchanging some of their individuals. On a computational grid, such operation guarantees that the EAs are at the same evolution stage, and so prevents the non-effect and super-individual quoted below. However, due the heterogeneity issue the synchronous mode is less performant in term of consumed CPU time. Indeed, the evolution process is often hanging on powerful machines waiting the less powerful ones to complete their computation. On the other hand, the synchronous island model is not fault tolerant as wasting an island implies the blocking of the whole model in a volatile environment. The synchronous mode is globally more complex and less performant on a computational grid.

The memory of the island model required for the checkpointing mechanism is composed of the memory of each of its EAs and the individuals being migrated. One has to note that in an asynchronous mode such memory is not critical as the EAs are stochastic processes.

6.3.2 The Parallel Evaluation of the Population

The evaluation of the population is time intensive and often represents more than 90% of the CPU time consumed by an EA especially for real-world problems. Its parallelization is thus necessary to reduce the execution time without changing the semantics of the EA in comparison to a serial execution. The parallel evaluation is often performed according to the farmer/worker model (see Figure 6.2). The farmer applies the selection, transformation and replacement operations as they require a global management of the population. At each generation, it distributes the new solutions among the workers, which evaluate them and return back their corresponding fitness values. The granularity (number of evaluators) has a great impact on the efficiency of the parallel model. This is the only parameter the user may possibly provide. The model is generic and could be provided in a software platform in a transparent way for the user.

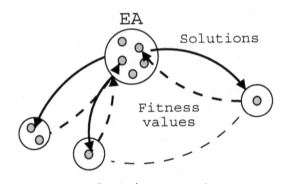

Fig. 6.2. The parallel evaluation of a population

In the multi-objective context, the implementation approach often used consists in distributing the population among the different workers. Each worker evaluates the vector of all the objective functions on all the received individuals. Another approach consists in distributing the different objective functions among the workers, and each of them computes the value of its assigned function on each individual of the population. The farmer will then aggregate the partial results for all the individuals of the population. Such approach allows a degree of parallelism and extendibility limited to the number of objective functions, meaning often 2 or 3.

According to the order in which the evaluation phase is performed in comparison with the other parts of the EA, two modes can be distinguished: *synchronous* and *asynchronous*. In the synchronous mode, the *farmer* manages the evolution process and performs in a serial way the different steps of selection, transformation and replacement. At each iteration, it distributes the set of new generated solutions among the workers and waits for the results to be returned back. After the results are collected, the evolution process is re-started. The model does not change the semantics of the EA compared to a serial execution.

The synchronous execution of the model is always synchronized with the return back of the last evaluated solution. Therefore, the model is blocking and thus less performant on an heterogeneous computational grid. Moreover, as the model is not fault tolerant the disappearance of an evaluating agent requires the redistribution of its individuals to other agents. As a consequence, it is essential to store all the solutions not yet evaluated. From another point of view, the scalability of the model is limited to the size of the population.

In the asynchronous mode, the evaluation phase is not synchronized with the other parts of the EA. The farmer does not wait for the return back of all the evaluations to perform the selection, transformation and replacement steps. The *steady-state* EA is a good example illustrating the asynchronous model and its several advantages. It is naturally non-blocking and fault tolerant to such a degree a threshold of wasted evaluations is not exceeded because any control is performed on the solutions sent for evaluation. In addition, the asynchronous mode is well adapted in term of efficiency to applications with an irregular evaluation cost of the objective function. Furthermore, it is not necessary to store the solutions evaluated by the workers, and the memory usage is thus optimized. On the other hand, the scalability is not limited by the size of the population.

6.3.3 The Parallel Evaluation of a Single Solution

The fitness of each solution is evaluated in a parallel centralized way (see Figure 6.3). This kind of parallelism could be efficient if the evaluation function is CPU time-consuming and/or IO intensive. This is often particularly true for industrial optimization problems. The function could be viewed as an aggregation of a set of partial functions. A reduction operation is performed on the results returned back by the partial functions computed by the workers. Consequently, for this model the user has to provide a set of partial functions and their associated aggregation operator. The user has thus to be skilled a little bit in parallel computing. In the multi-objective context, these partial functions could be the different objectives to be optimized.

The model is important for solving real-world problems for which the objective function is time intensive. For instance, the function may require the access to a huge database that could not be managed on a single machine. Due to a material constraint, the database is distributed among different sites, and data parallelism is exploited in the evaluation of the objective function. The parallelization of the objective function is also interesting when the objective function is CPU time consuming. Although its scalability is limited its utilization together with the parallel (particularly synchronous) evaluation of the population allows to extend the scalability of this latter. In the

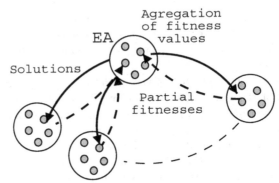

Fig. 6.3. The parallel evaluation of a single solution

multi-objective context, the scalability could be improved again if the different objective functions are simultaneously parallelized.

The parallel evaluation of a single solution model requires new problem-specific features mainly the vector of partial functions and an associated aggregation function. The implementation of the model being always synchronous, it is necessary to store in the memory associated to the model the values of the different partial solutions to manage the fault tolerant and dynamic availability of the grid.

The model is often neglected in the taxonomies on parallel EAs proposed in the literature [17]. However, its exploitation jointly with the other parallel models allows to significantly improve the degree of parallelism especially in the synchronous mode. Indeed, such result has been proved in our research works on the mobile network design problem [16]. It consists in positioning base stations on potential sites in order to fulfill some objectives and constraints [18]. More exactly, the multi-objective problem is to find a set of sites for antennas from a set of pre-defined candidates locations, to determine the type and the number of antennas, and their configuration parameters (tilt, azimuth, power, ...). Exactly three objectives are considered: (1) Maximizing the communication traffic; (2) Minimizing the total number of sites; and, (3) Minimizing the communications interferences. On some instances of the problem, the degree of parallelism associated with the parallel evaluation of a single solution (a network design) achieves 20. Therefore, if one considers an island model with 10 islands each of them with 100 individuals taking into account the parallel evaluation of a single solution model extends the degree of parallelism (number of simultaneous evaluations) from 1000 to 20.000.

6.4 Frameworks for Grid-based EAs – The ParadisEO-CMW Case

Several frameworks for the reusable design of parallel and distributed EAs have been proposed and most of them are available on the Web. The major of them are the following: DREAM[2] [19], ECJ[3] [20], JDEAL[4] [21] and Distributed BEAGLE[5] [22]. These frameworks are reusable as they are based on a clear object-oriented conceptual separation between solution methods and optimization problems. They are also portable as they are developed in Java except the latter, which is coded in C++. However, they are limited regarding the parallel distributed models. Indeed, in DREAM and ECJ only the island model is implemented using Java threads and TCP/IP sockets. JDEAL provides only the Master-Slave parallel evaluation of the population model using TCP/IP sockets. Distributed BEAGLE provides this latter model and the synchronous migration-based island model. Except DREAM, all these cited frameworks do not allow the deployment of the provided algorithms on computational grids or peer-to-peer systems. The Distributed Resource Evolutionary Algorithm Machine (DREAM) Project [19] (http://www.world-wide-dream.org) is a peer-to-peer software infrastructure devoted to support info-habitants evolving in an open and scalable way. It considers a virtual pool of distributed computing resources, where the different steps of an EA are automatically and transparently processed. However, only the island model is provided and this model does not need a great amount of resources. In [2], we have proposed a framework called *ParadisEO* dedicated to the reusable design of parallel and distributed EAs for only dedicated parallel hardware platforms. In [3], we have extended it to deal with computational grids.

6.4.1 ParadisEO: an Extended EO

The "EO" part of ParadisEO means Evolving Objects because it is basically an extension of the Evolving Objects (EO) [23] framework. EO is an LGPL C++ open source framework downloadable from http://eodev.source forge.net. The framework is originally the result of an European joint work [23]. EO includes a paradigm-free Evolutionary Computation library (EOlib) dedicated to the flexible design of EAs through evolving objects superseding

---

[2] Distributed Resource Evolutionary Algorithm Machine: http://www.world-wide-dream.org

[3] Java Evolutionary Computation: http://www.cs.umd.edu/projects/plus/ec/ecj/

[4] Java Distributed Evolutionary Algorithms Library: http://laseeb.isr.ist.utl.pt/sw/jdeal/

[5] Distributed Beagle Engine Advanced Genetic Learning Environment: http://www.gel.ulaval.ca/beagle

the most common dialects (Genetic Algorithms, Evolution Strategies, Evolutionary Programming and Genetic Programming). Flexibility is enabled through the use of the object-oriented technology. Templates are used to model the EA features: coding structures, transformation operators, stopping criteria, etc. These templates can be instantiated by the user according to his/her problem-dependent parameters. The object-oriented mechanisms such as inheritance, polymorphism, etc. are powerful ways to design new algorithms or evolve existing ones. Furthermore, EO integrates several services making it easier to use, including visualization facilities, on-line definition of parameters, application checkpointing, etc.

In its original version, EO does not provide any facility for parallel computing. Sticking out such limitation was the main objective in the design and development of ParadisEO. The ParadisEO framework [2] is dedicated to the reusable design of parallel hybrid meta-heuristics by providing a broad range of features including EAs, local search methods, parallel and distributed models, different hybridization mechanisms, etc. ParadisEO is a C++ LGPL extensible open source framework based on a clear conceptual separation of the meta-heuristics from the problems they are intended to solve. This separation and the large variety of implemented optimization features allow a maximum code and design reuse. Such conceptual separation is expressed at implementation level by splitting the classes in two categories: *provided classes* and *required classes*. The provided classes constitute a hierarchy of classes implementing the invariant part of the code. Expert users can extend the framework by inheritance/specialization. The required classes coding the problem-specific part are abstract classes that have to be specialized and implemented by the user. The classes of the framework are fine-grained, and instantiated as evolving objects embodying each one only one method. This is a particular design choice adopted in EO and ParadisEO. The heavy use of these small-size classes allows more independence and thus a higher flexibility compared to other frameworks. Changing existing components and adding new ones can be easily done without impacting the rest of the application.

ParadisEO is one of the rare frameworks that provide the most common parallel and distributed models described above. These models are portable as well on distributed-memory machines as on shared-memory multi-processors as they are implemented using standard libraries such as MPI, PVM and PThreads. The models can be exploited in a transparent way, one has just to instantiate their associated ParadisEO components. The user has the possibility to choose by a simple instantiation MPI or PVM for the communication layer. The models have been validated on academic and industrial problems, and the experimental results demonstrate their efficiency [2].

6.4.2 ParadisEO-CMW: Grid-enabled ParadisEO

The first release of ParadisEO allows a transparent exploitation of the parallel models on dedicated environments. In [3], the focus is on their re-design and deployment on large-scale non-dedicated computational grids. This is a great challenge as nowadays there is no effective grid-enabled framework for meta-heuristics to the best of our knowledge. To Grid-enable ParadisEO, first of all, one needs a Grid middleware and two interfaces: an infrastructure interface and a Grid Application Programming Interface. The first one provides communication and resource management tools. Our approach consists in using the Condor high-throughput computing system [4] as a Grid infrastructure. In addition, the MW abstract programming framework [5] supplies the two required interfaces. The global architecture of ParadisEO-CMW is layered as it is illustrated in Figure 6.4.

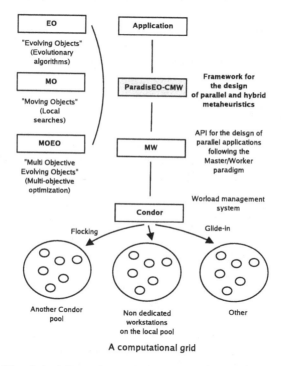

Fig. 6.4. A layered architecture of ParadisEO-CMW

From top to down, the first level supplies the optimization problems to be solved using the framework. The second level represents the ParadisEO framework including optimization solvers embedding single and multi-objective meta-heuristics (evolutionary algorithms and local searches). The third level provides interfaces for Grid-enabled programming and for access to the Condor infrastructure. The fourth and lowest level supplies communication and resource management services.

One of the major issues to deal with in the gridification process of ParadisEO is the ability of the framework to allow deployments on multi-administrative domains. Such service is provided in Condor by the flocking mechanism. Another important issue to deal with is fault-tolerance (volatility characteristic). MW automatically reschedules on Worker Nodes unfinished tasks which were running on processors that failed. However, this can not be applied to the master process that launches and controls tasks on Worker Nodes. Nevertheless, a couple of primitives are provided by MW to fold up or unfold the whole application enabling the user to save/restart the state in/from a file stream. Note that only the Master Node is concerned by these checkpointing operations. Dealing with EAs, these functionalities are easily investigated and checkpointing EAs is straightforward. It consists in saving/restoring the memory content associated to each parallel model defined in Section 6.3. In ParadisEO-CMW, default checkpointing policies are initially associated to the deployed meta-heuristics, and can be exploited by the users in a transparent way. In addition, these policies can be extend with more application-specific features.

6.5 Conclusion

In this chapter, we have presented an analysis of parallel EAs. The analysis is a contribution in parallel evolutionary computation as unlike previously published studies it is placed within the context of grid computing. Furthermore, such work is a part of a methodological approach for the reusable design of parallel EAs transparently deployable on computational grids. The objective is to visit again the traditional parallel models in order to allow their adaptation to the grids. Four major characteristics of grids have been considered: volatility, heterogeneity, large scale and multi-administrative domain.

The volatility issue requires the design and implementation of a fault tolerance approach in particular for synchronous EAs. Such mechanism is not always provided in the grid middlewares, and when it is provided it is based on the re-starting of interrupted work units from scratch. This solution is not efficient for applications in which work units are time-consuming such as those generated by the parallel models of EAs. A better solution is to propose as it is done in this chapter a checkpointing approach that consists in associating to each model a memory that determines the current state of parallel evolution process.

The heterogeneity nature of the grid sets the non-effect and/or super-individual problem for the island model and the efficiency problem for the two other models. The first problem could be solved by migrating islands from powerful machines to poor ones and conversely. To solve the efficiency problem of farmer-worker models it is necessary to implement them according to a pull or cycle stealing mode. In this way, more powerful processors will naturally perform more evaluations and the global load is well balanced.

In a large scale environment, it is important to deploy all the parallel models in a hierarchical way. The parallel evaluation of a single solution often neglected in the literature is very important to improve the degree of parallelism. The scheduling of the different work units of the different models is crucial for efficiency. For instance, in a multi-administrative domain grid the evaluations generated by each island must be allocated if possible to the processors belonging to the same domain. In addition, in such grid the security issue could be solved with different approaches. For instance, one could use a middleware [12] allowing large scale cycle stealing that naturally bypasses firewalls. The GSI service provided by Globus [11] is also a good solution.

The development of EAs using parallel computing and grid technologies is often complex and painful. Therefore, frameworks allowing a transparent use of these advanced features are very important for the evolutionary computation community. However, few of such frameworks exist and these are often limited in terms of provided models. In this chapter, we have presented ParadisEO-CMW [3] as a complete case study. This framework is an LGPL C++ open source software that provides the implementation of the three models on computational grids. These models have been experimented and validated on several academic and real-world applications such as the attribute selection in spectroscopic data mining [3].

References

1. I. Foster and C. Kesselman. *The Grid: Blueprint for a New Computing Infrastructure*, chapter Chapter 2: Computational Grids. Morgan-Kaufman, San Francisco, CA, 1999.
2. S. Cahon, N. Melab, and E-G. Talbi. ParadisEO: A Framework for the Reusable Design of Parallel and Distributed Metaheuristics. *Journal of Heuristics*, Vol. 10:353–376, May 2004. Kluwer Academic Publishers http://www.lifl.fr/~cahon/paradisEO/.
3. S. Cahon, N. Melab, and E-G. Talbi. An Enabling Framework for Parallel Optimization on the Computational Grid. In the Proc. of the 5th IEEE/ACM Intl. Symp. on Cluster Computing and the Grid (*CCGRID'2005*), Cardiff, UK, 9-12 May, 2005.
4. Miron Livny, Jim Basney, Rajesh Raman, and Todd Tannenbaum. Mechanisms for High Throughput Computing. *SPEEDUP Journal*, 11(1), June 1997. http://www.cs.wisc.edu/condor/.
5. J. Linderoth, S. Kulkarni, J.P. Goux, and M. Yoder. An Enabling Framework for Master-Worker Applications on the Computational Grid. In *Proc. of the 9th IEEE Symposium on High Performance Distributed Computing (HPDC9)*, pages 43–50, Pittsburgh, PA, Aug. 2000. http://www.cs.wisc.edu/condor/mw/.
6. J.H. Holland. *Adaptation in natural and artificial systems*. Ann Arbor, MI, USA, The University of Michigan Press, 1975.
7. I. Foster, C. Kesselman, and S. Tuecke. The anatomy of the Grid: Enabling Scalable Virtual Organizations. *Int. J. High Perform. Comput. Appl.*, 15(3):200–222, 2001.

8. K. Krauter, R. Buyya, and M. Maheswaran. A taxonomy and survey of grid resource management systems for distributed computing. *Software - Practice & Experience*, 32(2):135–164, 2002.
9. Miguel L. Bote-Lorenzo, Yannis A. Dimitriadis, and Eduardo Gómez-Sánchez. Grid Characteristics and Uses: A Grid Definition. In *European Across Grids Conference, LNCS 2970*, Lecture Notes in Computer Science, pages 291–298, 2003.
10. Ian Foster. What is the Grid? A Three Point Checklist. *Grid Today*, 1(6), July 22 2002.
11. I. Foster and C. Kesselman. Globus: A Metacomputing Infrastructure Toolkit. *Intl. J. of Supercomputer Applications*, 11(2):115–128, 1997.
12. Gilles Fedak. *XtremWeb: une plate-forme pour l'étude expérimentale du calcul global pair-à-pair*. PhD thesis, Université Paris XI, 2003.
13. J.P. Cohoon, S.U. Hedge, W.N. Martin, and D. Richards. Punctuated equilibria: A parallel genetic algorithm. In *Proc. of the Second Intl. Conf. on Genetic Algorithms*, page 148. Lawrence Erlbaum Associates, 1987.
14. T. Starkweather, D. Whitley, and K. Mathias. Optimization using distributed genetic algorithms. In H.-P. Schwefel and R. Manner, editors, *Parallel Problem Solving from Nature*, Volume 496, page 176. LNCS, Springer-Verlag, 1991.
15. T. Belding. The distributed genetic algorithm revisited. In D. Eshelmann editor, editor, *Sixth Int. Conf. on Genetic Algorithms*, San Mateo, CA, 1995. Morgan Kaufmann.
16. N. Melab. Contributions la résolution de problèmes d'optimisation combinatoire sur grilles de calcul. Thèse d'Habilitation à Diriger des Recherches (HDR), Université de Lille1, Novembre, 2005.
17. E. Cantú-Paz. A Survey of Parallel Genetic Algorithms, 1997.
18. H. Meunier, El-Ghazali Talbi, and P. Reininger. A Multiobjective Genetic Algorithm for Radio Network Optimization. In *Congress on Evolutionary Computation*, Volume 1, pages 317–324. IEEE Service Center, July 2000.
19. M.G. Arenas, P. Collet, A.E. Eiben, M. Jelasity, J.J. Merelo, B. Paechter, M. Preuss, and M. Schoenauer. A framework for distributed evolutionary algorithms. In *Proceedings of PPSN VII*, september 2002.
20. S. Luke, L. Panait, J. Bassett, R. Hubley, C. Balan, and A. Chicop. ECJ: A Java-based Evolutionary Computation and Genetic Programming Research system, 2002. http://www.cs.umd.edu/projects/plus/ec/ecj/.
21. J. Costa, N. Lopes, and P. Silva. Jdeal: The Java Distributed Evolutionary Algorithms Library. 2000.
22. C. Gagné, M. Parizeau, and M. Dubreuil. Distributed BEAGLE: An Environment for Parallel and Distributed Evolutionary Computations. In *Proc. of the 17th Annual Intl. Symp. on High Performance Computing Systems and Applications (HPCS) 2003*, May 11–14 2003.
23. M. Keijzer, J.J. Morelo, G. Romero, and M. Schoenauer. Evolving Objects: A General Purpose Evolutionary Computation Library. *Proc. of the 5th Intl. Conf. on Artificial Evolution (EA'01)*, Le Creusot, France, Oct. 2001.

7

Parallel Evolutionary Algorithms on Consumer-Level Graphics Processing Unit

Tien-Tsin Wong[1] and Man Leung Wong[2]

[1] Department of Computer Science and Engineering
   The Chinese University of Hong Kong, Shatin, Hong Kong
   ttwong@cse.cuhk.edu.hk
[2] Department of Computing and Decision Sciences
   Lingnan University, Tuen Mun, Hong Kong
   mlwong@ln.edu.hk

Evolutionary Algorithms (EAs) are effective and robust methods for solving many practical problems such as feature selection, electrical circuits synthesis, and data mining. However, they may execute for a long time for some difficult problems, because several fitness evaluations must be performed. A promising approach to overcome this limitation is to parallelize these algorithms. In this chapter, we propose to implement a parallel EA on consumer-level *Graphics Processing Unit* (GPU). We perform experiments to compare our parallel EA with an ordinary EA and demonstrate that the former is much more effective than the latter. Since consumer-level graphics processing units are already widely available and installed on oridinary personal computers and they are easy to use and manage, more people will be able to use our parallel algorithm to solve their problems encountered in real-world applications.

7.1 Introduction

Evolutionary Algorithms (EAs) are weak search and optimization techniques inspired by natural evolution [1]. They have been demonstrated to be effective and robust in searching very large and varied spaces in a wide range of applications such as feature selection [2], electrical circuits synthesis [3], and data mining [4, 5, 6, 7]. In general, EAs include all population-based algorithms that use selection and recombination operators to generate new search points in a search space. They include genetic algorithms [8, 9], genetic programming [10, 3, 11], evolutionary programming [12, 13], and evolution strategies [14, 15].

T. Wong and M.L. Wong: *Parallel Evolutionary Algorithms on Consumer-Level Graphics Processing Unit*, Studies in Computational Intelligence (SCI) **22**, 133–155 (2006)
www.springerlink.com   © Springer-Verlag Berlin Heidelberg 2006

The various kinds of EAs differ mainly in the evolution models applied, the evolutionary operators employed, the selection methods, and the fitness functions used [12]. Genetic Algorithms (GAs) and Genetic Programming (GP) model evolution at the level of genetic. They emphasize the acquisition of genetic structures at the symbolic level and regularities of the solutions. On the other hand, the idea of optimization is used in Evolution Strategies (ES) and the structures being optimized are the individuals of the population. Various behavioral properties of the individuals are parameterized and their values evolved as an optimization process. Evolutionary Programming (EP) uses the highest level of abstraction by emphasizing the adaptation of behavioral properties of various species.

Although EAs are effective in solving many practical problems in science, engineering, and business domains, they may execute for a long time to find solutions for some huge problems, because several fitness evaluations must be performed. A promising approach to overcome this limitation is to parallelize these algorithms for parallel, distributed, and networked computers. However, these specialized computers are relatively more difficult to use, manage, and maintain. Moreover, some people may not have access to these specialized computers. On the other hand, powerful graphics processing units (GPU) have already been *widely* available and installed on oridinary personal computers, due to the popularity of computer games and its strong support by the dominant Microsoft development standard. Unlike the specialized parallel computers, the extra cost for using consumer-level GPUs is low. In fact, there will be effectively no "extra cost" because GPU will soon become a compulsory component on ordinary personal computers. Therefore, we propose to implement a parallel EA on consumer-level GPUs. Given the ease of use, maintenance, and management of GPUs and personal computers, more people will be able to use our parallel algorithm to solve huge problems encountered in real-world applications such as data mining.

In the following section, different parallel and distributed EAs will be described. Graphics processing unit will be discussed in Section 7.3. We will present our parallel evolutionary algorithm in Sections 7.4 and 7.5. A number of experiments have been performed and the experimental results will be discussed in Section 7.6. We will give a conclusion and a description of our future work in Section 7.7.

7.2 Parallel and Distributed Evolutionary Algorithms

For almost all practical applications of EAs, most computation time is consumed in evaluating the fitness value of each individual in the population since the genetic operators of EAs can be performed efficiently. Memory availability is another important problem of EAs because the population usually has a large number of individuals.

There is a relation between the difficulty of the problem to be solved and the size of the population. In order to solve substantial and real-world problems, a population size of thousands and a longer evolution process are usually required. A larger population and a longer evolution process imply more fitness evaluations must be conducted and more memory are required. In other words, a lot of computational resources are required to solve substantial and practical problems. Usually, this requirement cannot be fulfilled by normal workstations. Fortunately, these time-consuming fitness evaluations can be performed independently for each individual in the population and individuals in the population can be distributed among multiple computers.

EAs have a high degree of inherent parallelism which is one of the motivation of studies in this field. In natural populations, thousands or even millions of individuals exist in parallel and these individuals operates independently with a little cooperation and/or competition among them. This suggests a degree of parallelism that is directly proportional to the population size used in EAs. There are different ways of exploiting parallelism in EAs: master-slave models; improved-slave models; fine-grained models; island models; and hybrid models [16].

The most direct way to implement a parallel EA is to implement a global population in the master processor. The master sends each individual to a slave processor and let the slave to find the fitness value of the individual. After the fitness values of all individuals are obtained, the master processor selects some individuals from the population using some selection method, performs some genetic operations, and then creates a new population of offspring. The master sends each individual in the new population to a slave again and the above process is iterated until the termination criterion is satisfied.

Another direct way to implement a parallel EA is to implement a global population and use the tournament selection which approximates the behavior of ranking. Assume that the population size N is even and there are more than $N/2$ processors. Firstly, $N/2$ slave processors are selected. A processor selected from the remaining processors maintains the global population and controls the overall evolution process and the $N/2$ slave processors. Each slave processor performs two independent m-ary tournaments. In each tournament, m individuals are sampled randomly from the global population. These m individuals are evaluated in the slave processor and the winner is kept. Since there are two tournaments, the two winners produced can be crossed in the slave processor to generate two offspring. The slave processor may perform further modifications to the offspring. The offspring are then sent back to the global population and the master processor proceeds to the next generation if all offspring are received from the $N/2$ slave processors.

Fine-grained EAs explore the computing power of massively parallel computers such as the Maspar. To explore the power of this kind of computers, one can assign one individual to each processor, and allow each individual to seek a mate close to it. A global random mating scheme is inappropriate because of the limitation of the communication abilities of these computers.

Each processor can select probabilistically an individual in its neighborhood to mate with. The selection is based on the fitness proportionate selection, the ranking, the tournament selection, and other selection methods proposed in the literature. Only one offspring is produced and becomes the new resident at that processor. The common property of different massively parallel evolutionary algorithms is that selection and mating are typically restricted to a local neighborhood.

Island models can fully explore the computing power of course grain parallel computers. Assume that we have 20 high performance processors and have a population of 4000 individuals. We can divide the total population into 20 subpopulations (islands or demes) of 200 individuals each. Each processor can then execute a normal evolutionary algorithm on one of these subpopulations. Occasionally, the subpopulations would swap a few individuals. This migration allows subpopulations to share genetic material. Since there are 20 independent evolutionary searches occur concurrently, these searches will be different to a certain extent because the initial subpopulations will impose a certain sampling bias. Moreover, genetic drift will tend to drive these subpopulations in different directions. By employing migration, island models are able to exploit differences in the various subpopulations. These differences maintain genetic diversity of the whole population and thus can prevent the problem of premature convergence.

Hybrid models combine several parallelization approaches. The complexity of these models depends on the level of hybridization.

7.3 Graphics Processing Unit

In the last decade, the need from the multimedia and games industries for accelerating 3D rendering has driven several graphics hardware companies devoted to the development of high-performance parallel graphics accelerator. This results in the birth of GPU, which is designed to accelerate standardized rendering requests. The whole pipeline consists of the transformation, texturing, illumination, and rasterization to the framebuffer. The need for cinematic rendering from the games industry further raised the need for programmability of the rendering process. Starting from the recent generation of GPUs launched in 2001 (including nVidia GeforceFX series and ATI Radeon 9800 and above), developers can write their own C-like programs, which are called *shaders*, on GPU. Due to the wide availability, programmability, and high-performance of these consumer-level GPUs, they are cost-effective for, not just game playing, but also scientific computing.

Unfortunately, GPUs are tailored for graphics programming, not general-purpose computation. We need to work around the rendering pipeline in order to adapt the system for scientific computing. For instance, users have to define dummy triangles to fill up the whole screen in order to trigger the shader for computation. In the rest of this section, we first describe the general procedure in the graphics pipeline. Then, a simple example is used to illustrate the typical procedure in order to utilize the GPU for scientific computing.

There are two major types of shaders (vertex shader and fragment shader) corresponding to the two major modules of the rendering pipeline, namely vertex and fragment engines. The user first defines the triangles by specifying the 3D position of each vertex. It seems irrelevant to define 3D triangles for evolutionary computation. However, such declaration is necessary for satisfying the input format of the graphics pipeline. The texture coordinate associating with each vertex is also defined at the same time. These texture coordinates are needed to define the correspondence of elements in textures (input/output data buffers) and the pixels on the screen (shaders are executed on per-pixel basis). For each vertex, a vertex shader (user-defined program) is executed (Fig. 7.1). The shader program must be *Single-Instruction-Multiple-Data* (SIMD) in nature, i.e. the same set of operations has to be executed on different vertices. The triangles are then projected onto the 2D screen and rasterized (discretized) into many fragments (pixels) in the framebuffer as shown in Fig. 7.1.

Next, the fragment engine takes place. For each pixel, one or more user-defined fragment shaders are executed to process data associated with this pixel. In the shader program, the input textures can be fetched for computation and results are output via the output textures. Texture is basically a data buffer. Similarly, the fragment shader must also be SIMD in nature. Due to the design of GPUs, it is not allowed to explicitly allocate pixels, and hence the jobs, to different processors. Logically, shader programs are executed on each pixel independently and in parallel.

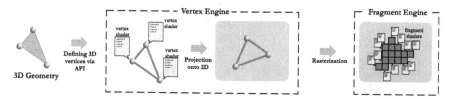

Fig. 7.1. The 3D rendering pipeline

As an example of utilizing GPU for scientific computing, we illustrate the addition of two $M \times N$ matrices, P and Q. Although this example is simple, it demonstrates the typical procedure in setting up the GPU for many scientific computing applications. Firstly, we define two right triangles (one upper and one lower) covering the $M \times N$ screen pixels as shown on the left hand side of Fig. 7.2. The vertex shader basically does nothing but only projects the six vertices (of two triangles) onto the 2D screen. After rasterization, these two triangles are broken down into $M \times N$ fragments (or pixels). For each pixel, a fragment shader is executed. Normally, the fragment shader is more useful in scientific computing than the vertex shader.

We then fed matrices P and Q to this shader via two *input textures*, each of size $M \times N$ (Fig. 7.2). Most data has to be input to the parallel program (shader) via the form of textures. The correspondence between the texture elements and the screen pixels is usually one-to-one and defined via the texture coordinate at each vertex. A texture is basically an image (2D array) with each pixel composed of four components, (r, g, b, α). Readers can regard each pixel as a 4D vector. Each component can hold a 32-bit floating point value. Therefore, one way to add two matrices is to store P's elements in the r component of one input texture and Q's elements in the r component of another texture. Obviously, a more compact and practical representation is to store elements of P and Q in two components, say r and b, of the same texture. For presentation clarity, we use two input textures. As the fragment shader is executed at each pixel (x, y) *independently* and *in parallel*, it only contains one single addition statement and no looping is needed (Fig. 7.2). The statement fetches and sums $P(x, y).r$ and $Q(x, y).r$, and stores the output in the third texture, $O(x, y).r$. The notation $.r$ specifies the r component of the pixel. The high performance is mainly contributed by this SIMD-type parallelism. Most GPU nowadays impose certain limitations on the usage of textures. For example, the total number of textures being accessed simultaneously is usually limited (e.g. 16 textures on nVidia GeForceFX 6800). Furthermore, the input texture cannot be used for output.

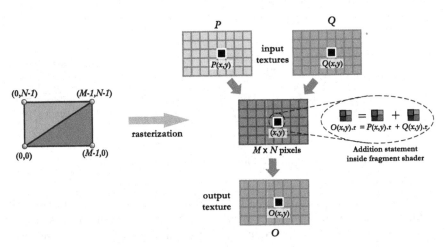

Fig. 7.2. Addition of two matrices on GPU

7.4 Data Organization

Suppose we have μ individuals and each contains k variables (genes). The most natural representation for an individual is an array. As GPU is tailored for parallel processing and optimized multi-channel texture fetching, all input data to GPU should be loaded in the form of *textures*. Fig. 7.3 shows how we represent μ individuals in form of texture. Without loss of generality, we take $k = 32$ as an example of illustration throughout this paper. This amount of variables reflects the typical size of real-world problems.

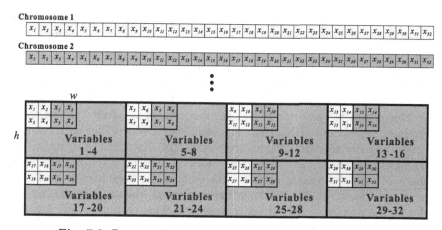

Fig. 7.3. Representing individuals of 32 variables on textures

As each pixel in the texture contains *quadruple* of 32-bit floating point values (r, g, b, α), we can encode an individual of 32 variables into 8 pixels. In other words, the memory is more efficiently utilized if k is multiple of 4. This is also why we take $k = 8 \times 4 = 32$ as a working example. Instead of mapping an individual to 8 consecutive pixels in the texture, we divide an individual into quadruple of 4 variables. The same quadruples from all individuals are grouped and form a tile in the texture as shown in Fig. 7.3. Each tile is $w \times h = \mu$ in size. The reason we do not adopt the *consecutive-pixel* representation is that the implementation will be complicated when k varies. Imagine the complication of variables' offsets within the texture when k increases from 32 to 48. On the other hand, the *fragmentation-and-tiling* representation is more scalable because increasing k can be easily achieved by adding more tiles. Another reason is for the sake of visualization. We shall explain the details in Section 7.6.2. In our specific example of $k = 32$, 4×2 tiles are formed. It is up to user to decide the organization of these tiles in the texture. The first tile (upper-left tile) in Fig. 7.3 stores variables 1 to 4, while the next tile stores variables 5 to 8, and so on.

Texture on GPU is not as flexible as main memory. Current GPUs impose several limitations. One of them is the size of texture must not exceed certain limit, e.g. 4096×4096 on nVidia GeforceFX 6800. In other words, to fit the whole population in one texture on our GPU, we must satisfy $k\mu \leq 4 \times 4096^2$. For extremely large populations with a large number of variables, multiple textures have to be used. Note that there are also limitation on the total number of textures that can be accessed simultaneously. The actual number varies on different GPU models. Normally, at least 16 textures can be supported.

7.5 Evolutionary Programming on GPU

Evolutionary programming (EP) and genetic algorithm (GA) have been both successfully applied to several numerical and optimization problems. While classical GA requires the processes of crossover and mutation, EP requires the mutation process only. Hence, for each generation of evolution, EP is less computational intensive than GA. When implementing on GPU, the crossover process of GA induces more *rendering passes* than that of EP.

One complete execution of the fragment shader is referred as one rendering pass. On current GPU, there is a significant overhead for each rendering pass. The more rendering passes are needed, the slower the program is. Since fragment shaders are executed independently on each pixel, no information sharing is allowed among pixels. If the computation result of a pixel A has to be used for computing an equation at pixel B, the computation result of A must be written to an output texture first. This output texture has to be fed to the shader for computation in *next* rendering pass. Therefore, if the problem being tackled involves a chain of data dependency, more rendering passes are needed, and hence the speed-up is decreased.

Since the crossover process of GA requires more passes and more data transfer than that of EP, EP is more *GPU-friendly* (efficient to implement on GPU) than GA. Hence, in this paper, we study the GPU implementation of EP instead of the classical GA. Without loss of generality, we assume the optimization is to minimize a cost function. Hence, our EP is used to determine a $\boldsymbol{x}_{\min}$, such that

$$\forall \boldsymbol{x}, f(\boldsymbol{x}_{\min}) \leq f(\boldsymbol{x})$$

where $\boldsymbol{x} = \{x_i(1), x_i(2), \ldots, x_i(k)\}$ is the individual containing k variables; $f : R^n \mapsto R$ is the function being optimized. We implement a fast evolutionary programming (FEP) based on Cauchy mutation [17] as shown in Table 7.1.

In the algorithm, $\boldsymbol{x}_i$ is the individual evolving and $\boldsymbol{\eta}_i$ controls the vigorousness of mutation of $\boldsymbol{x}_i$. In general, the computation of FEP can be roughly divided into three types: (a) mutation and reproduction (step 3), (b) fitness value evaluation (steps 2 and 4), and (c) competition and selection (steps 5 and 6). These types of operations will be discussed in the following sub-sections.

Table 7.1. The Fast Evolutionary Programming Algorithm

1. Generate the initial population of μ individuals, each of which can be represented as a set of real vectors, $(x_i, \eta_i), i = 1, \ldots, \mu$. Both x_i and η_i contain k independent variables,
$$x_i = \{x_i(1), \ldots, x_i(k)\}$$
$$\eta_i = \{\eta_i(1), \ldots, \eta_i(k)\}$$
2. Evaluate the fitness score for each individual (x_i, η_i), $i = 1, \ldots, \mu$, of the population based on the objective function, $f(x)$.
3. For each parent (x_i, η_i), $i = 1, \ldots, \mu$, create an offspring (x_i', η_i') as follows: for $j = 1, \ldots, k$
$$x_i'(j) = x_i(j) + \eta_i(j)R(0, 1),$$
$$\eta_i'(j) = \eta(j) \exp(\tfrac{1}{\sqrt{2k}} R(0, 1) + \tfrac{1}{\sqrt{2\sqrt{k}}} R_j(0, 1))$$
where $x_i(j), \eta_i(j), x_i'(j)$, and $\eta_i'(j)$ denote the j-th component of x_i, η_i, x_i', and η_i' respectively. $R(0, 1)$ denotes a normally distributed 1D random number with zero mean and standard deviation of one. $R_j(0,1)$ indicates a new random variable for each value of j.
4. Calculate the fitness of each offspring (x_i', η_i').
5. Conduct pairwise comparison over the union of parents (x_i, η_i) and offspring (x_i', η_i'), for $i = 1, \ldots, \mu$. For each individual, q (tournament size) opponents are chosen randomly from all the parents and offspring. For each comparison, if the individual's fitness is smaller than or equal to that of opponent, it receives a "**win**".
6. Select μ individuals out of (x_i, η_i) and (x_i', η_i'), $i = 1, \ldots, \mu$, that receive more **win**'s to be parents of next generation.
7. Stop if the stopping criterion is satisfied; otherwise go to Step 3.

7.5.1 Mutation and Reproduction

Unlike GA, EP omits crossover and carries out mutation only. Fogel [18] introduced EP using Gaussian distribution. Yao and Liu [17] proposed a mutation operation based on Cauchy distribution to increase the speed of convergence. From the pseudocode above, mutation operation is executed on each variable. Variables are assumed to be independent of each other. Thus mutation process is perfectly parallelizable. In pure software (CPU for short) implementation, a loop is needed to perform mutation on each variable. On the SIMD-based GPU, a fragment shader is executed *in parallel* to perform mutation on each component (r, g, b, α) of each pixel. GPU solution is thus ideal for this independent mutation and can achieve significant speed-up.

To accomplish the mutation process on GPU, we designed two fragment shaders, one for computing x' and the other for η'. Fig. 7.4 illustrates these two shaders graphically. The parents x_i and η_i are stored in two input textures while the offspring are generated and written to two output textures x_i' and η_i'. One fragment shader is responsible for computing x_i' while the other

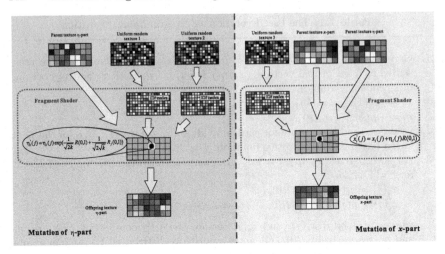

Fig. 7.4. The two fragment shaders for mutation process

is responsible for η_i'. Besides, we also need two input textures of random numbers.

Mutation requires normally distributed random variables. Unfortunately, current GPU is not equipped with random number generator. Hence the random numbers have to be generated by CPU and fed to GPU in the form of input textures. We divide the process of random number generation into two steps. Firstly, CPU is used to generate random variables with uniform distribution. This is a sequential process. The generated random numbers are fed to GPU via input textures. Then, inside the two fragment shaders, GPU converts them from uniform distribution to Gaussian distribution in parallel.

Traditionally, the well-known Box-Muller transformation [19] is used to transform random numbers of uniformly distribution to normal distribution. The polar form of Box-Muller transformation algorithm provides even faster and more robust solution [20]. However, the number of iterations in Box-Muller transformation depends on the data values. Such data-dependent looping is undesirable for SIMD-based GPU, as various processors execute different numbers of iterations.

Instead of using Box-Muller transformation, we employ the direct inverse cumulative normal distribution function (ICDF) as it does not require looping. It trades accuracy for speed. The algorithm uses minimax approximation and the error introduced is relatively little. Our experiment showed that GPU implementation of ICDF is 2 times faster than CPU implementation of ICDF. ICDF on GPU is even more than 4 times faster than Box-Muller transformation on CPU. We also found that the transformation does not affect the quality of the solutions generated by the evolutionary algorithms.

Current GPU has a slow performance in data transferal from GPU texture to main memory. Therefore, such data transfer should be avoided as much as possible. Hence, our strategy is to keep the parents and offspring in GPU memory. Only the final result, after several generations of evolution, is transferred from GPU textures to main memory.

7.5.2 Fitness Value Evaluation

Fitness value evaluation determines the "goodness" of individuals. It is one of the core part of EP. After each evolution, the fitness value of each individual in the current population is calculated. The result is then passed to the later stage of EP process. Each individual returns a fitness value by feeding the objective function f with the values of the variables of the individual. This evaluation process usually consumes most of the computational time.

Since no interaction between individuals is required during evaluation, the evaluation is fully parallelizable. Fig. 7.5 illustrates the evaluation shader graphically. Recall that the individuals are broken down into quadruples and stored in the tiles within the textures. The evaluation shader hence looks up the corresponding quadruple in each tile during the evaluation. The fitness values are output to an output texture of size $w \times h$, instead of $4w \times 2h$, because each individual only returns a single value.

7.5.3 Competition and Selection

Replacing the old generation is the last stage of each evolution. There are two major processes involved, competition and selection. EP employs a stochastic selection (soft selection) through the tournament schema. Each individual in the union set of parent and offspring population takes part in a q-round tournament. In each round, an opponent is randomly drawn from the union set of parents and offsprings. The number of opponents defeated is recorded by the variable win. After the tournament, a selection process takes place and chooses the best μ individuals having the highest win values as parents for next generation.

Competition

Exploitation is the process of using information gathered to determine which searching direction is profitable. In EP, exploitation is realized by competition. Individuals in the population compete with q randomly drawn opponents, where q is the predefined tournament parameter. The considered individual wins if its fitness value is better than (in our case smaller than or equal to) that of the opponent. The times of winning is recorded in win. It tells us how good this individual is.

Competition can be done either on GPU or CPU. For GPU implementation, q textures of random values have to be generated by CPU and loaded to GPU memory for each evolution. On the other hand, for CPU implementation, only the fitness textures of both parent and offspring population have

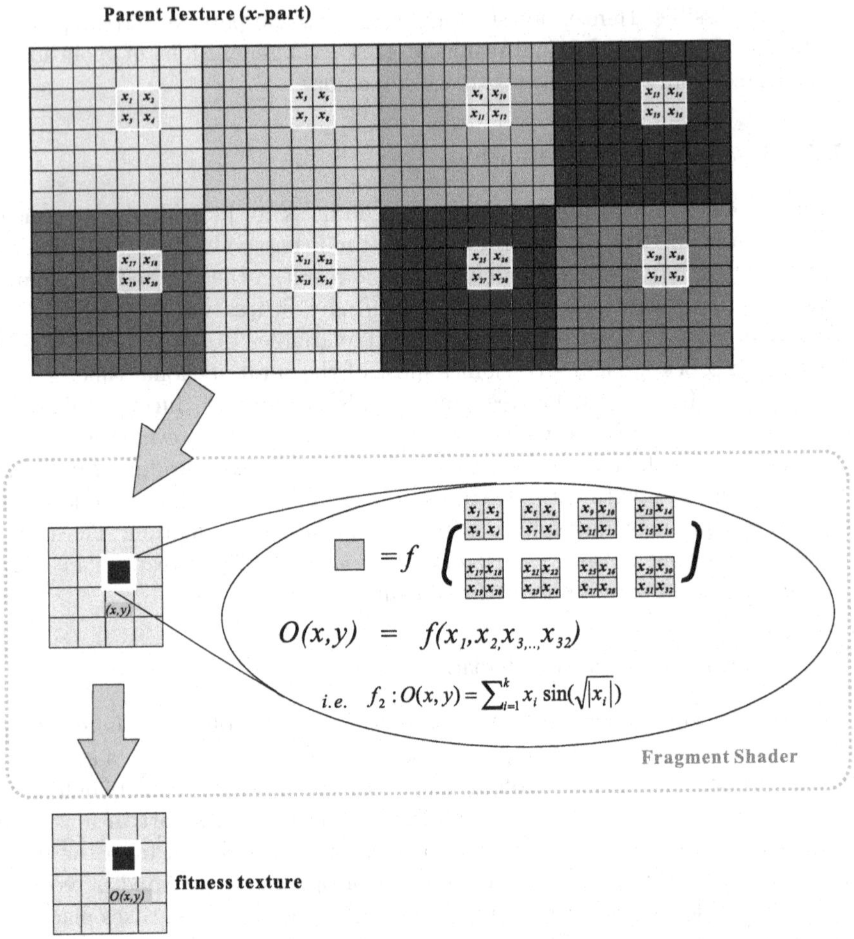

Parent Texture (x-part)

Fig. 7.5. The shader for fitness evaluation

to be transferred from GPU memory to main memory. The final result of selected individuals is then transferred back to GPU memory.

It seems that GPU implementation should be faster than the CPU one, as it parallelizes the competition. However, our experiments show that using GPU to implement the competition is slower than that of using CPU. The major bottleneck of GPU implementation is the transfer of q textures towards GPU memory. It evidences the limitation of slow data transfer of current GPU. As the competition does not involve any time-consuming fitness evaluation (it only involves fitness comparison), the gain of parallelization does not compensate the loss due to data transfer. The data transfer rate is expected to be significantly improved in the future GPU.

Median Searching and Selection

After the competition process, selection is performed based on the win values. It selects the best μ individuals having highest win values and assigns them as the parents for the next generation. The most natural way is to sort the 2μ individuals in a descending order of win values. The first μ individuals are then selected. For large population size, the sorting time is unbearably slow even using $O(N \log(N))$ sorting algorithm.

Note that our goal is to pick the best μ individuals, instead of sorting the individuals. These two goals are different. We can pick the best μ individuals without sorting them if we know the median win value. The trick is to find the median without any sorting.

Median searching has been well studied. Floyd and Rivest [21] proposed a complex partition-and-conquer linear time algorithm. When the valid range of values is known and countable, median searching can be done easily. To illustrate the idea, we go through a running example in Fig. 7.6. Suppose we have $\mu = 9$ values (top of Fig. 7.6) and the valid range of value is an integer within $[0, q]$ where $q = 5$ in our example. Now, we construct $q + 1 = 6$ bins as shown in the middle of Fig. 7.6. Therefore, in a linear time, we scan through all 9 values and form a histogram showing the count of each valid value (bin). Next we compute the cumulative distribution function (bottom part of Fig. 7.6) based on the histogram. The median is the value whose cumulated sum exceeds half the number of elements, i.e. $\mu/2 = 4.5$. In our example, value 3 is the median. Both the construction of histogram and cumulative distributed function are linear in time complexity. Therefore, we can find the median in linear time without sorting. Once the median is known, we select all individuals with fitness values smaller than the median. Some individuals with fitness values equal to the median will also be chosen until μ individuals are picked.

Minimizing Data Transfer

To minimize the data transfer between the memory on GPU and the main memory, an index array storing the offset of each individual in the textures is constructed before the competition and selection process. During the selection, we only record the index of the selected individuals, instead of the whole individuals. The index array is then loaded to GPU in form of a texture. The actual individual replacement is performed on GPU based on the index texture. The final result is rendered to a texture which stores the individuals of the new generation. With this approach, the textures of individuals are always retained in the GPU memory. These textures of individuals are never transferred to the main memory during the evolution, until the final generation is obtained. Only fitness textures, random textures and index textures are transferred between GPU and CPU during the evolution. Since the fitness, random and index textures are smaller than the textures of individuals,

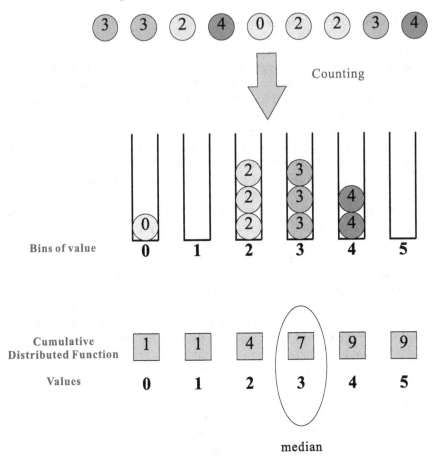

Fig. 7.6. Running example of median picking algorithm

this indexing approach minimizes the data transfer and improves the speed significantly.

7.6 Experimental Results and Visualization

7.6.1 Experimental Results

We applied EP with Cauchy distribution to a set of benchmark optimization problems. Table 7.2 summarizes the benchmark functions, number of variables and the search ranges. We conducted the experiments for 20 trials on both CPU and GPU. The average performance is reported in this paper. The experiment test bed was an Pentium IV 2.4 GHz with AGP 4X enabled

Table 7.2. The set of test functions. N is the number of variables. S indicates the ranges of the variables and f_m is the minimum value of the function

Test Functions	N	S	f_m
$f_1 : \sum_{i=1}^{N} x_i^2$	32	$(-100, 100)^N$	0
$f_2 : \sum_{i=1}^{N} (\sum_{j=1}^{i} x_j)^2$	32	$(-100, 100)^N$	0
$f_3 : \sum_{i=1}^{N-1} \{100(x_{i+1} - x_i^2)^2 + (x_i - 1)^2\}$	32	$(-30, 30)^N$	0
$f_4 : -\sum_{i=1}^{N} x_i \sin(\sqrt{|x_i|})$	32	$(-500, 500)^N$	-12569.5
$f_5 : \sum_{i=1}^{N} \{x_i^2 - 10 \cos(2\pi x_i) + 10\}$	32	$(-5.12, 5.12)^N$	0

consumer-level GeForce 6800 Ultra display card, with 512 MB main memory and 256 MB GPU memory. The following parameters were used in the experiments:

- population size: $\mu = 400, 800, 3200, 6400$
- tournament size: $q = 10$
- standard deviation: $\sigma = 1.0$
- maximum number of generation: $G = 2000$

Fig. 7.7 for functions f_1 and f_2, Fig. 7.8 for functions f_3 and f_4 and Fig. 7.9 for function f_5 show, by generation, the average fitness value of the best solutions found by the CPU and GPU approaches with various population sizes in 20 trials. Since the two approaches generate the same solutions, a curve is depicted for each population size. It can be observed that better solutions can be obtained for all functions if a larger population size is used. This phenomenon can be explained because more search points are available in a larger population and EP is less likely to be trapped in local minimum. For the population size of 6400, the average fitness values of the best solutions found at the last generation are respectively 0.003, 1349.88, 52.88, −13259.3, and 2.41 for f_1, f_2, f_3, f_4, and f_5. The minimum values f_m listed in Table 7.2 can be achieved if the algorithms are executed for more generations.

However, EP with a larger population size will take longer execution time. Fig. 7.10 for functions f_1 and f_2, Fig. 7.11 for functions f_3 and f_3 and Fig. 7.12 for function f_5 display, by generation, the average execution time of the GPU and CPU approaches with different population sizes. ¿From the curves in this figure, the execution time increases if a larger population is applied. However, our GPU approach is much more efficient than the CPU implementation because the execution time of the former is much less than that of the latter if the population size reaches 800. Moreover, the efficiency leap becomes larger when the population size increases.

The ratios of the average execution time of the GPU (CPU) approach with population sizes of 800, 3200, and 6400 to that of the corresponding approach with population size of 400 are summarized in Table 7.3. It is interesting to notice that, the CPU approach shows a linear relation between the number of

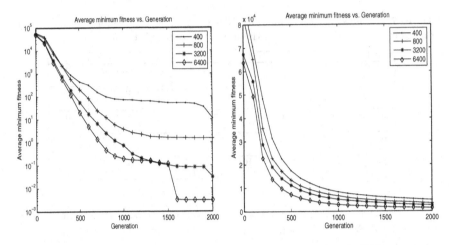

Fig. 7.7. Fitness value of the best solution found by the GPU and CPU approaches for functions f_1 and f_2. The results were averaged over 20 independent trials

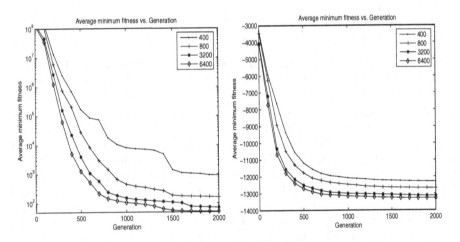

Fig. 7.8. Fitness value of the best solution found by the GPU and CPU approaches for functions f_3 and f_4. The results were averaged over 20 independent trials

individuals and the execution time, while our GPU approach has a sub-linear relation. For example, our GPU approach with population sizes of 400 and 800 take about the same execution time. Moreover, the execution time of our approach with population size of 6400 is about 3 times of that with population size of 400. Definitely, this is an advantage when huge population sizes are required in some real-life applications.

To study why our approach can achieve this phenomenon, the average execution time of different types of operations of the GPU (CPU) approach for the test function f_5 are presented in Table 7.4. It can be observed that

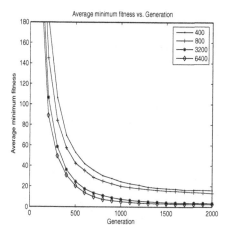

Fig. 7.9. Fitness value of the best solution found by the GPU and CPU approaches for function f_5. The results were averaged over 20 independent trials

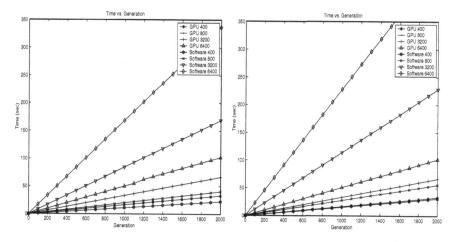

Fig. 7.10. Execution time of the GPU and CPU approaches for functions f_1 and f_2. The results were averaged over 20 independent trials

the fitness evaluation time of our GPU approach with different population sizes are about the same, because all individuals are evaluated in parallel. Moreover, the mutation time does not increase proportionally with the number of individuals, because the mutation operations are also executed in parallel [3]. Similar results are also obtained for other test functions. Table 7.5 displays the speed-ups of our GPU approach with the CPU approach. The speed-ups depend on the population size and the problem complexity. Generally, GPU

---

[3] The mutation time increases with the number of individuals, because our GPU approach requires a number of random numbers generated by CPU.

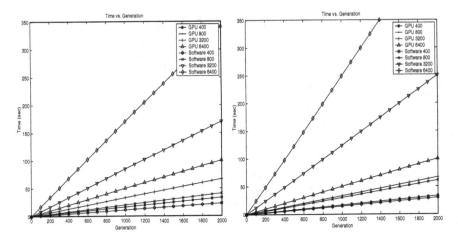

Fig. 7.11. Execution time of the GPU and CPU approaches for functions f_3 and f_4. The results were averaged over 20 independent trials

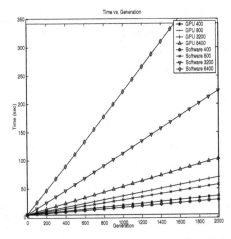

Fig. 7.12. Execution time of the GPU and CPU approaches for function f_5. The results were averaged over 20 independent trials

outperforms CPU when the population size is larger than or equal to 800. The speed-up ranges from about 1.25 to about 5.02. For complicated problems that require huge population sizes, we expect that GPU can achieve even better performance gain.

Fig. 7.13 for functions f_1 and f_2, Fig. 7.14 for functions f_3 and f_4 and Fig. 7.15 for function f_5 display the relation between the average fitness value of the best solutions and the average execution time. These curves can be interpreted as how much time is needed to achieve the average fitness value. In general, EP with small population sizes can find good solutions in short

Table 7.3. The ratios of the average execution time of the GPU (CPU) approach with different population sizes to that with population size of 400

μ	GPU					CPU				
	f_1	f_2	f_3	f_4	f_5	f_1	f_2	f_3	f_4	f_5
800	1.00	1.00	1.00	1.00	1.00	2.01	2.02	2.02	2.02	2.02
3200	2.02	2.02	2.02	2.02	2.02	8.30	8.24	8.37	8.12	8.29
6400	3.11	3.09	3.04	3.05	3.05	16.57	16.45	16.75	16.40	16.53

Table 7.4. Experimental result summary of f_5

μ	type	competition & selection time (sec)	speed-up	fitness eval-uation time (sec)	speed-up	mutation time (sec)	speed-up	total time (sec)	speed-up
400	CPU	3.32	0.80	7.28	0.28	16.19	7.39	26.79	0,82
	GPU	4.14		26.46		2.19		32.79	
800	CPU	6.70	0.91	14.86	0.68	32.49	9.00	54.05	1.65
	GPU	7.33		21.84		3.61		32.78	
3200	CPU	28.26	1.06	60.25	2.31	133.52	9.92	222.03	3.35
	GPU	26.72		26.12		13.46		66.30	
6400	CPU	56.69	1.08	118.91	5.41	267.30	10.44	442.95	4.42
	GPU	52.57		21.96		25.61		100.25	

Table 7.5. The speed-up of the GPU approach

μ	f_1	f_2	f_3	f_4	f_5
400	0.62	0.85	0.62	0.93	0.82
800	1.25	1.71	1.25	1.88	1.65
3200	2.55	3.45	2.57	3.74	3.35
6400	3.31	4.50	3.42	5.02	4.42

runs. However, for long runs, EP with large population sizes always guarantee better solutions while EP with small population sizes will saturate at some sub-optimal positions. From the curves, GPU outperforms CPU at long runs (i.e. around 60 seconds). The execution time of GPU with large population sizes is approximately the same as that of CPU with small population sizes, while the former can find better solutions.

7.6.2 Visualization

As GPU is designed for display, all on-board textures can be trivially visual-ized in real-time without much additional cost. Such visualization allows users to instantly observe individuals of the current generation. We designed two visualization schemes, namely *genome-convergence* and *fitness-convergence*.

Recall that individuals are broken down into quadruples of variables (Fig. 7.3) and stored in (r, g, b, α). By mapping the minimum and maximum values of a variable to $[0, 255]$ (8-bit integer), we can regard the values in each quadruple as a color and output to the screen. In other words, we visu-alize the genome map (Fig. 7.3). At the beginning of the evolution, the values are basically random and hence displayed as a noise image in Fig. 7.16(a). As the population converges, the color of each tile becomes less noisy and

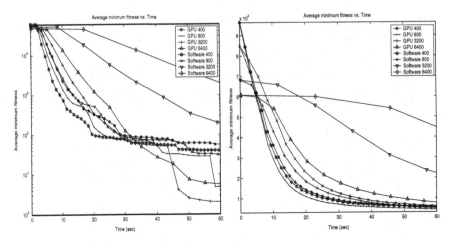

Fig. 7.13. Fitness time graphs for functions f_1 and f_2. The results were averaged over 20 independent trials

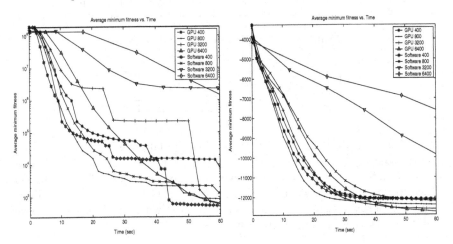

Fig. 7.14. Fitness time graphs for functions f_3 and f_4. The results were averaged over 20 independent trials

converges to a single color. In fact, the actual color of each tile does not matter. The most important observation is the apparentness of boundaries between two consecutive tiles, as different variables may converge to different values. Fig. 7.16(a)-(d) show the genome-convergence map at iterations 100, 500, 1000, and 2000 respectively (**The boundaries are more apparent when observed on screen with colors**). Therefore, a useful indicator of convergence is the apparentness of tile boundaries.

The fitness-convergence visualization scheme is more traditional. During our fitness evaluation, we obtain a texture of fitness values. This texture can

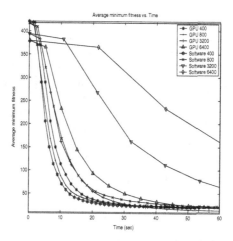

Fig. 7.15. Fitness time graphs for function f_5. The results were averaged over 20 independent trials

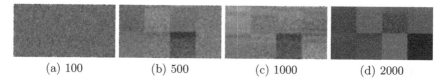

(a) 100   (b) 500   (c) 1000   (d) 2000

Fig. 7.16. Four snapshots of genome-convergence visualization

be output to screen for inspection. Again, we can map the fitness values to a range of color values for better visualization using simple shader program. As our test functions are all minimization problems, we map the minimum (0) and maximum fitness values to $[0, 255]$ gray levels for visualization. Fig. 7.17 shows 4 snapshots of fitness values during the evolution. Similarly, the first snapshot shows the randomness (noisiness) when the evolution begins. As the evolution continues, the fitness image converges to a less noisy image.

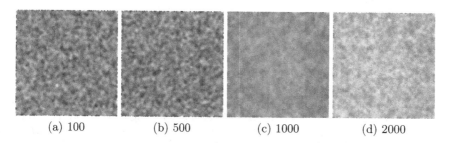

(a) 100   (b) 500   (c) 1000   (d) 2000

Fig. 7.17. Four snapshots of fitness-convergence visualization

7.7 Summary

In this research, we have implemented a parallel EP on consumer-level graphics processing units and proposed indirect indexing and many optimization skills to achieve maximal efficiency. The parallel EP is a hybrid of master-slave and fine-grained models. Competition and selection are performed by CPU (i.e. the master) while fitness evaluation, mutation, and reproduction are performed by GPU which is essentially a massively parallel machine with shared memory. Unlike other fine-grained parallel computers such as Maspar, GPU allows processors to communicate with, not only nearby processors, but also any other processors. Hence more flexible fine-grained EAs can be implemented on GPU. We have done experiments to compare our parallel EP on GPU and an ordinary EP on CPU. It is found that the speed-up factor of our parallel EP ranges from 1.25 to 5.02, when the population size is large enough. Moreover, there is a sub-linear relation between the population size and the execution time. Thus, our parallel EP will be very useful for solving difficult problems that require huge population sizes.

For future work, we plan to implement a parallel genetic algorithm on GPU and compare it with the approach reported in this paper.

Acknowledgment

This work is supported by The Chinese University of Hong Kong Young Researcher Award (Project No. 4411110) and the Earmarked Grant LU 3009/02E from the Research Grant Council of the Hong Kong Special Administrative Region.

References

1. P. Angeline, "Genetic programming and emergent intelligent," in *Advances in Genetic Programming*, Jr K. E. Kinnear, Ed., pp. 75–97. MIT Press, Cambridge, MA, 1994.
2. Il-Seok Oh, Jin-Seon Lee, and Byung-Ro Moon, "Hybrid genetic algorithms for feature selection," *IEEE Transactions on Pattern Analysis and Machine Intelligence*, vol. 26, no. 11, pp. 1424–1437, 2004.
3. John R. Koza, M. A. Keane, M. J. Streeter, W. Mydlowec, J. Yu, and G. Lanza, *Genetic Programming IV: Routine Human-Competitive Machine Intelligence*, Kluwer Academic Publishers, 2003.
4. M. L. Wong, W. Lam, and K. S. Leung, "Using evolutionary programming and minimum description length principle for data mining of Bayesian networks," *IEEE Transactions on Pattern Analysis and Machine Intelligence*, vol. 21, no. 2, pp. 174–178, February 1999.
5. M. L. Wong, W. Lam, K. S. Leung, P. S. Ngan, and J. C. Y. Cheng, "Discovering knowledge from medical databases using evolutionary algorithms," *IEEE Engineering in Medicine and Biology Magazine*, vol. 19, no. 4, pp. 45–55, 2000.

6. Man Leung Wong and Kwong Sak Leung, "An efficient data mining method for learning Bayesian networks using an evolutionary algorithm based hybrid approach," *IEEE Transactions on Evolutionary Computation*, vol. 8, no. 4, pp. 378–404, 2004.

7. Alex A. Freitas, *Data Mining and Knowledge Discovery with Evolutionary Algorithms*, Springer, 2002.

8. J. H. Holland, *Adaptation in Natural and Artificial Systems*, University of Michigan Press, 1975.

9. David E. Goldberg, *Genetic Algorithms in Search, Optimization, and Machine Learning*, Addison-Wesley, 1989.

10. John R. Koza, *Genetic Programming: on the Programming of Computers by Means of Natural Selection*, MIT Press, 1992.

11. Wolfgang Banzhaf, Peter Nordin, Robert E. Keller, and Frank D.Francone, *Genetic Programming: An Introduction*, Morgan Kaufmann, San Francisco, California, 1998.

12. David B. Fogel, *Evolutionary Computation: Toward a New Philosohpy of Machine Intelligence*, IEEE Press, 2000.

13. L. Fogel, A. Owens, and M. Walsh, *Artificial Intelligence Through Simulated Evolution*, John Wiley and Sons, 1966.

14. H. P. Schewefel, *Numerical Optimization of Computer Models*, John Wiley and sons, New York, 1981.

15. Thomas Bäck, *Evolutionary Algorithms in Theory and Practice: Evolution Strategies, Evolutionary Programming, Genetic Agorithms*, Oxford University Press, 1996.

16. Erick Cantú-Paz, *Efficient and Accurate Parallel Genetic Algorithms*, Kluwer Academic Publishers, 2000.

17. X. Yao and Y. Liu, "Fast evolutionary programming," in *Evolutionary Programming V: Processdings of the 5th Annual Conference on Evolutionary Programming*. 1996, Cambridge, MA:MIT Press.

18. David B. Fogel, "An introduction to simulated evolutionary optimization," *IEEE Transactions on Neural Networks*, vol. 5, no. 1, pp. 3–14, 1994.

19. G. E. P. Box and M. E. Muller, "A note on the generation of random normal deviates," *Annals of Mathematical Statistics*, vol. 29, pp. 610–611, 1958.

20. D. E. Knuth, "The art of computer programming. volume 2: Seminumerial algorithms (second edition)," *Addison-Wesley, Menlo Park*, 1981.

21. Robert W. Floyd and Ronald L. Rivest, "Expected time bounds for selection," *Communications of the ACM*, vol. 18(3), pp. 165–172, 1975.

Part IV

Parallel Particle Swarm Optimization

8

Intelligent Parallel Particle Swarm Optimization Algorithms

Shu-Chuan Chu and Jeng-Shyang Pan

[1] Department of Information Management,
   Cheng Shiu University, Taiwan
   scchu@csu.edu.tw, http://home.mis.csu.edu.tw/scchu/
[2] Department of Electronic Engineering, National Kaohsiung University of
   Applied Sciences University,
   Kaohsiung City, Taiwan
   jspan@cc.kuas.edu.tw, http://bit.kuas.edu.tw/~ jspan/

Some social systems of natural species, such as flocks of birds and schools of fish, possess interesting collective behavior. In these systems, globally sophisticated behavior emerges from local, indirect communication amongst simple agents with only limited capabilities. In an attempt to simulate this flocking behavior by computers, Kennedy and Eberthart (1995) realized that an optimization problem can be formulated as that of a flock of birds flying across an area seeking a location with abundant food. This observation, together with some abstraction and modification techniques, led to the development of a novel optimization technique - particle swarm optimization.

Particle swarm optimization has been shown to be capable of optimizing hard mathematical problems in continuous or binary space. We present here a parallel version of the particle swarm optimization (PPSO) algorithm together with three communication strategies which can be used according to the independence of the data. Some communication strategies for PPSO are discussed in this work, which can be used according to the strength of the correlation of parameters. Experimental results confirm the superiority of the PPSO algorithms.

8.1 Introduction

Soft computing is an emerging collection of methodologies to exploit tolerance for uncertainty, imprecision and partial truth to achieve useable robustness, tractability, low total cost and approximate solutions. It is particularly efficient and effective for NP-hard problems. Recently, many differently challenges

S.-C. Chu and J.-S. Pan: *Intelligent Parallel Particle Swarm Optimization Algorithms*, Studies in Computational Intelligence (SCI) **22**, 159–175 (2006)
www.springerlink.com

posed by data mining have been solved by various soft computing methodologies. At this juncture, the main components of soft computing involve fuzzy theory, artificial neural networks, genetic algorithms [12, 6, 11], simulating annealing [16, 13], tabu search approach, swarm intelligence systems (such as particle swarm optimization, ant systems and ant colony systems [2, 3]) and other approaches related to cognitive modelling. Each of them contributes a revealable methodology which only in a cooperative rather than competitive manner for persuading problems in its field. The hybridization of soft computing is shown in Figure 8.1.

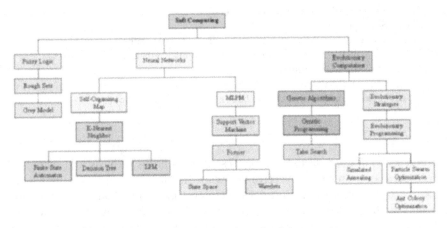

Fig. 8.1. The hybridization of soft computing

Some social systems of natural species, such as bird flock and fish school, possess interesting collective behavior. In these systems, globally sophisticated behavior emerges from local, indirect communication among simple agents with only limited capabilities. In an attempt to simulate this flocking behavior by computers, Kennedy and Eberthart (1995) realized that an optimization problem can be formulated as that of a flock of birds flying across an area seeking a location with abundant food. This observation, together with some abstraction and modification techniques, led to the development of a novel optimization technique – particle swarm optimization.

Unlike using genetic operator, particle swarm optimization is a population-based evolutionary computation technique [8, 14, 21] and has similarities to the general evolutionary algorithm. PSO optimizes an objective function by conducting population-based search. The population consists of potential solutions, called particles, which are metaphor of birds in bird flocking. The particle are randomly initialized and then freely fly across the multi-dimensional search space. During the flying, every particle updates its velocity and position based on the best experience of its own and the entire population. The updating policy will drive the particle swarm to move toward region with higher object value, and eventually all particles will gather around the point with highest object value. However, PSO is motivated from the simulation of social behavior which differs from the natural selection scheme of genetic algorithms [12, 6, 11]. One advantage of PSO is that it often locates near optima significantly faster than evolutionary optimization [1, 9].

PSO processes the search scheme using populations of particles which correspond to the use of individuals in genetic algorithms. Each particle is equivalent to a candidate solution of a problem. The particle will move according to an adjusted velocity, which is based on the corresponding particle's experience and the experience of its companions. For the D-dimensional function $f(.)$, the ith particle for the tth iteration can be represented as

$$X_i^t = (x_i^t(1), x_i^t(2), \ldots, x_i^t(D)). \tag{8.1}$$

Assume that the best previous position of the ith particle at the tth iteration is represented as

$$P_i^t = (p_i^t(1), p_i^t(2), \ldots, p_i^t(D)), \tag{8.2}$$

then

$$f(P_i^t) \leq f(P_i^{t-1}) \leq \cdots \leq f(P_i^1). \tag{8.3}$$

The velocity of the ith particle at the tth iteration can be represented as

$$V_i^t = (v_i^t(1), v_i^t(2), \ldots, v_i^t(D)). \tag{8.4}$$

$$G^t = (X^t(1), X^t(2), \ldots, X^t(D)) \tag{8.5}$$

is defined as the *best* position amongst all particles from the first iteration to the tth iteration, where *best* is defined by some function of the swarm. The original particle swarm optimization algorithm can be expressed as follows:

$$V_i^{t+1} = V_i^t + C_1.r_1.(P_i^t - X_i^t) + C_2.r_2.(G^t - X_i^t) \tag{8.6}$$

$$X_i^{t+1} = X_i^t + V_i^{t+1}, i = 0, \ldots N - 1 \tag{8.7}$$

where N is particle size, $-V_{max} \leq V_i^{t+1} \leq V_{max}$, ($V_{max}$ is the maximum velocity) and r_1 and r_2 are random numbers such that $0 \leq r_1, r_2 \leq 1$. $C_1.r_1.(P_i^t - X_i^t)$ is called the cognition component which only takes into account the particle's own experiences. The performance of the particle swarm optimization is based on the interaction of the individual particle and the group particles. The performance of the cognition component only is inferior due to no interaction between the different particles. The last term in the velocity update equation $C_2.r_2.(G^t - X_i^t)$, represents the social interaction between the particles.

The particle swarm optimization (PSO) is related to some evolutionary algorithms. Firstly, the PSO is a population-based algorithm which is similar to all evolutionary algorithms. Secondly, the historical personal best position is operated as part of the population is similar as the selection operator. Thirdly, the velocity update equation resembles the arithmetic crossover operator as in the evolutionary algorithm. Alternatively, the velocity update equation, without the V_i^t term, can be seen as a mutation operator. G^t offers a faster rate of convergence to exploit the local optimal information. This strategy maintains only a single best solution, called the global best particle amongst all the particles in the swarm. The global best particle acts as an attractor, pulling all the particles towards the best position across all particles. P_i^t tries to prevent premature convergence by maintaining the multiple attractors for exploring multiple areas.

A discrete binary version of the particle swarm optimization algorithm was also proposed by Kennedy and Eberhart [15]. The particle swarm optimization algorithm has been applied *inter alia* to optimize reactive power and voltage control [10] and human tremor [7]. A modified version of the particle swarm optimizer [18] and an adaption using the inertia weight, is parameter controlling the dynamic of flying, of the modified particle swarm[19] have also been presented. The latter version of the modified particle swarm optimizer can be expressed as

$$V_i^{t+1} = W^t.V_i^t + C_1.r_1.(P_i^t - X_i^t) + C_2.r_2.(G^t - X_i^t) \qquad (8.8)$$

$$X_i^{t+1} = X_i^t + V_i^{t+1}, i = 0, \dots N-1 \qquad (8.9)$$

where W^t is the inertia weight at the tth iteration, C_1 and C_2 are factors used to control the related weighting of corresponding terms. The weighting factors, C_1 and C_2, compromises the inevitable tradeoff between exploration and exploitation. Shi and Eberhart[20] have also applied fuzzy theory to adapt the particle swarm optimization algorithm. In addition, the explosion, stability and convergence of the PSO has been analyzed by Clerc and Kennedy [4].

The modified version of the particle swarm optimization algorithm with inertia weight can be depicted as follows:

Initialization: Generate N particles X_i^t, $i = 0, 1, ..., N-1$, N is the total particle size and t is the iteration number. Set $t = 1$

Evaluation: Evaluate the value of $f(X_i^t)$ for every particle.

Update: Update the velocity and particle positions using Eqs. 8.8 and 8.9.

Increment: Increase the iteration number t.

Termination: Repeat step 2 to 5 until the predefined value of the function is achieved or the maximum number of iterations has been reached. Record the best value of the function $f(G^t)$ and the best particle position G^t.

To experience the power of particle swarm optimization, applied program to the following test function, as visualized in Figure 8.2.

$$F(x,y) = -x sin(\sqrt{|x|}) - y sin(\sqrt{|y|}), \quad -500 < x, y < 500$$

where global optimum is at $F(-420.97, -420.97) = 837.97$.

In the tests above, both learning factors, C_1 and C_2, are set to a value of 2, and a variable inertia weight w is used according to the suggestion from

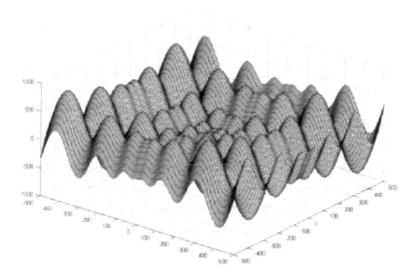

Fig. 8.2. Object function F

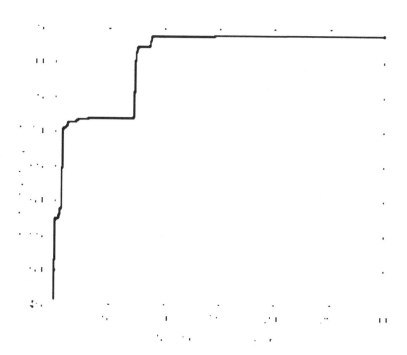

Fig. 8.3. Progress of PSO on object function F

Shi and Eberhart (1999). Figure 8.3 reports the progress of particle swarm optimization on the test function $F(x, y)$ for the first 300 iterations. At the end of 1000 iterations, $F(-420.97, -420.96) = 837.97$ is located, which is close to the global optimum.

It is worthwhile to look into the dynamics of particle swarm optimization. Figure 8.4 presents the distribution of particles at different iterations. There is a clear trend that particles start from their initial positions and fly toward the global optimum.

In this work, the concept of parallel processing to particle swarm optimization is exploited and a *Parallel Particle Swarm Optimization* (PPSO) idea is presented based on different solution space. Parallel PSO with communication strategies, which we termed PPSO, will be used to run the experiments presented in the following of this chapter.

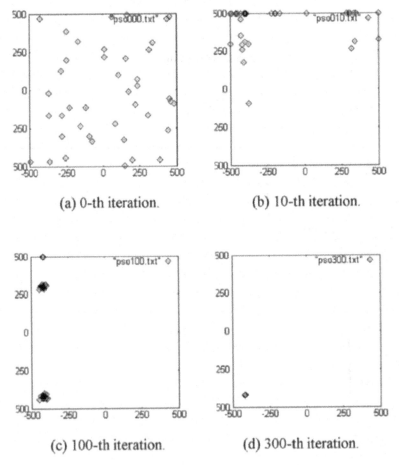

Fig. 8.4. The distribution of particles at different iterations

8.2 Parallel Particle Swarm Optimization

A conventional computer uses one processor which executes a set of instructions in order to produce results. At any instant time, there is only one operation being carried out by the processor. Parallel processing is concerned with producing the same results using multiple processors with the goal of reducing the running time. The goal of using parallel processing is to reduce the running time in a computer system. The two most common parallel processing methods are pipeline processing and data parallelism. The principle of pipeline processing is to separate the problem into a cascade of tasks where each of the tasks is executed by an individual processor as shown in Figure 8.5.

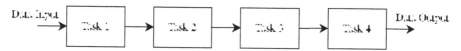

Fig. 8.5. Task and data distribution of pipeline processing

Data are transmitted through each processor which executes a different part of the process on each of the data elements. Since the program is distributed over the processors in the pipeline and the data moves from one processor to the next, no processor can proceed until the previous processor has finished its task. Data parallelism, as shown in Figure 8.6, is an alternative approach which involves distributing the data to be processed amongst all processors which then executes the same procedure on each subset of the data. Data parallelism has been applied fairly widely including to genetic algorithms.

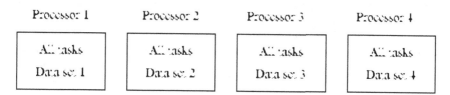

Fig. 8.6. Task and data distribution of data parallelism

The parallel genetic algorithm (PGA) works by dividing the population into several groups and running the same algorithm over each group using different processors [5]. However, the purpose of applying parallel processing to genetic algorithms goes further than merely being a hardware accelerator. Rather a distributed formulation is developed which gives better solutions with lower overall computation. In order to achieve this, a level of communication between the groups is performed every fixed number of generations. That is, the parallel genetic algorithm periodically selects promising individuals from each subpopulation and migrates them to different subpopulations.

With this migration (communication), each subpopulation will receive some new and promising chromosomes to replace the poorer chromosomes in a subpopulation. This strategy helps to avoid premature convergence. The parallel genetic algorithm has been successfully applied to vector quantization based communication via noisy channels [17]. In this work, the spirit of the data parallelism method is utilised to create a parallel particle swarm optimization (PPSO) algorithm [2].

It is difficult to find an algorithm which is efficient and effective for all types of problem. As shown in previous work by Shi and Eberhart [20], the fuzzy adaptive particle swarm optimization algorithm is effective for solutions which are independent or are loosely correlated such as the generalized Rastrigrin or Rosenbrock functions. However, it is not effective when solutions are highly correlated such as for the Griewank function. Our research has indicated that the performance of the PPSO can be highly dependent on the level of correlation between parameters and the nature of the communication strategy - this can be explained as follows.

Assuming the parameters of solutions are independent or only loosely correlated. For example, the parameters x_1 and x_2 for the minimization of $f(x_1, x_2) = x_1 + x_2$ are independent because we may keep x_2 constant and tune x_1 to minimize this function, then keep x_1 constant to tune x_2 for the minimization of this function, which is equivalent to tune both x_1 and x_2 simultaneously for minimizing $f(x_1, x_2)$. This means we may get the best value $x_1 = 0$ by keeping x_2 constant and vice versa. It does not loose the spirit for obtaining the global optima. So, if we tune the value of one parameter to get a better solution cost by keeping the other parameters constant, the value of this parameter is always in the neighborhood of the best solution. Based on this observation, we may update the best particle among all particles to each group and mutate to replace the poorer particles in each group very frequently. However, the above observation is not true if the parameters of solutions are strongly correlated. In fact the best solutions can be spread throughout the search space. In this case, we need to keep the parameters be divergent and the best particles cannot be used to replace the poorer particles for all groups and the communication frequency should be infrequent. Thus in this case it is more effective to limit the communication to the neighborhood only in order to retain the divergence. If the properties of the parameters are unknown, we may apply the communication strategy 3 which is the hybrid version of the communication strategy 1 and 2. Three communication strategies are thus presented and experiments have been carried out which show the utility of each strategy.

The mathematical form of the parallel particle swarm optimization algorithm can be expressed as follows:

$$V_{i,j}^{t+1} = W^t.V_{i,j}^t + C_1.r_1.(P_{i,j}^t - X_{i,j}^t) + C_2.r_2.(G_j^t - X_i^t) \qquad (8.10)$$

$$X_{i,j}^{t+1} = X_{i,j}^t + V_{i,j}^{t+1} \qquad (8.11)$$

$$f(G^t) \leq f(G^t_j) \tag{8.12}$$

where $i = 0, \ldots N_j - 1, j = 0, \ldots S - 1$, S $(= 2^m)$ is the number of groups (and m is a positive integer), N_j is the particle size for the jth group, $X^t_{i,j}$ is the ith particle position in the jth group at the tth iteration, $V^t_{i,j}$ is the velocity of the ith particle in the jth group at the tth iteration, G^t_j is the best position among all particles of the jth group from the first iteration to the tth iteration and G^t is the best position among all particles in all groups from the first iteration to the tth iteration.

As discussed above, three communication strategies have been developed for PPSO. The first strategy, shown in Figure 8.7, is based on the observation that if parameters are independent or are only loosely correlated, then the better particles are likely to obtain good results quite quickly. Thus multiple copies of the best particles for all groups G^t are mutated and those mutated particles migrate and replace the poorer particles in the other groups every R_1 iterations.

However, if the parameters of a solution are loosely correlated the better particles in each group tend not to obtain optimum results particularly quickly. In this case, a second communication strategy may be applied as depicted in Figure 8.8. This strategy is based on self-adjustment in each group. The best particle in each group G^t_j is migrated to its neighbour groups to replace some of the more poorly performing particles every R_2 iterations. Since we have defined the number of clusters S as a power of two, *neighbours* are defined as being those clusters where the binary representation of the cluster number j differs by one bit. In fact there are also some possible communication strategies between neighbor groups, such as updating the information between each pair groups and the pair group can be changed for each iteration shown in Figure 8.9. It is also possible to update the information according to the ring structure shown in Figure 8.10 or some other structures.

When the correlation property of the solution space is known the first and second communication strategies work well. However, they can perform poorly if applied in the wrong situation. As a result, in the cases where the correlation property is unknown, a hybrid communication strategy (i.e. strategy 3) can be applied. This hybrid strategy separates the groups into two equal sized subgroups with the first subgroup applying the first strategy every R_1 iterations and all groups applying the second strategy every R_2 iterations as depicted in Figure 8.11.

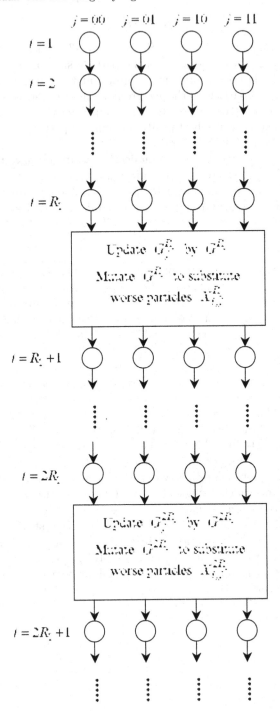

Fig. 8.7. Communication Strategy 1 for Loosely Correlated Parameters

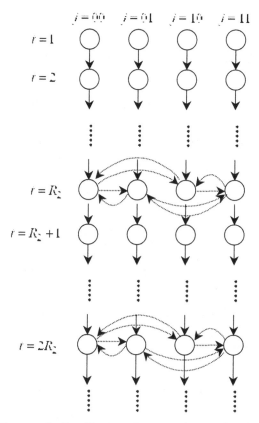

Fig. 8.8. Communication Strategy for strongly correlated parameters

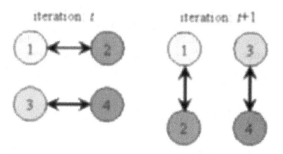

Fig. 8.9. Migrate best particle between each pair group

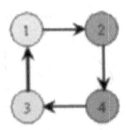

Fig. 8.10. Migrate the best particle according to the ring structure

---

Algorithm 8.1 Parallel Particle Swarm Optimization Algorithm

The complete parallel particle swarm optimization (PPSO) algorithm with its three communication strategies is as follows:

1. **Initialization:** Generate N_j particles $X_{i,j}^t$ for the jth group, $i = 0, \ldots N_j - 1, j = 0, \ldots S - 1$, S is the number of groups, N_j is the particle size for the jth group and t is the iteration number. Set $t = 1$.
2. **Evaluation:** The value of $f(X_{i,j}^t)$ for every particle in each group is evaluated.
3. **Update:** Update the velocity and particle positions using equations (8.10), (8.11) and (8.12).
4. **Communication:** Three possible communication strategies are as follows:

    Strategy 1: Migrate the best particle among all particles G^t to each group and mutate G^t to replace the poorer particles in each group and update G_j^t with G^t for each group, every R_1 iterations.

    Strategy 2: Migrate the best particle position G_j^t of the jth group to its neighbouring groups to substitute for some poorer particles, every R_2 iterations.

    Strategy 3: Separate the groups into two subgroups. Apply communication strategy 1 to subgroup 1 every R_1 iterations and communication strategy 2 to both subgroup 1 and subgroup 2 for every R_2 iterations.
5. **Termination:** Steps 2 to 5 are repeated until the predefined value of the function or some maximum number of iterations has been reached. Record the best value of the function $f(G^t)$ and the best particle position among all particles G^t.

8.3 Experimental Results

Let $X = \{x_1, x_2, \ldots, x_n\}$ be an n-dimensional real-value vector. The Rosenbrock function can be expressed as follows:

$$f_1(X) = \sum_{i=1}^{n} (100(x_{i+1} - x_i^2)^2 + (x_i - 1)^2).   \tag{8.13}$$

The second function used in the experiments was the generalized Rastrigrin function which can be expressed as:

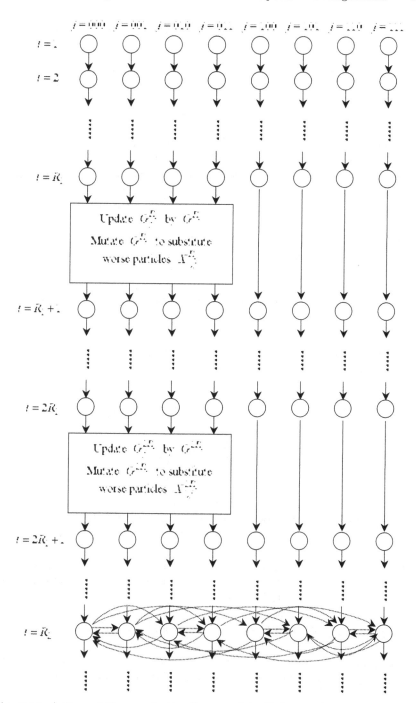

Fig. 8.11. A General Communication Strategy 3 for Unknown Correlation Between Parameters

$$f_2(X) = \sum_{i=1}^{n}(x_i^2 - 10cos(2\pi x_i)^2 + 10). \qquad (8.14)$$

The third function used was the generalized Griewank function as follows:

$$f_3(X) = \frac{1}{400}\sum_{i=1}^{n}x_i^2 - \prod_{i=1}^{n}cos\left(\frac{x_i}{\sqrt{i}}\right) + 1. \qquad (8.15)$$

Experiments were carried out to test the performance of the PPSO communication strategies. They confirmed that the first strategy works best when the parameters of the solution are loosely correlated such as for the Rastrigrin and Rosenbrock functions while the second strategy applies when the parameters of the solution are more strongly correlated as is the case for the Griewank function. A final experiment tested the performance of the third strategy for all three functions. All three experiments are compared with the linearly decreasing inertia weight PSO [19] for 50 runs.

The parameters of the functions for *PSO* and *PPSO* were set as in Table 8.1. We did not limit the value of X, C_1 and C_2 were set to 2, the maximum number of iterations was 2000, $W_t^0 = 0.9$, $W_t^{2000} = 0.4$ and the number of dimensions was set to 30.

Table 8.1. Asymmetric initialization ranges and V_{max} values

Function	Asymmetric Initialization Range	V_{max}
f_1	$15 \le x_i \le 30$	100
f_2	$2.56 \le x_i \le 5.12$	10
f_3	$300 \le x_i \le 600$	600

To ensure a fair comparison, the number of groups × the number of particles per group was kept constant – the particle size for the *PSO* was 160, one swarm with 160 particles, as reported by 1 × 160. For *PPSO*, the particle size was also set to be 160 that was divided into 8 groups with 20 particles in each group (i.e. 8 × 20), 4 groups with 40 particles in each group (i.e. 4 × 40) and 2 groups with 80 particles in each group (i.e. 2 × 80), respectively. For the first experiment, the number of iterations for communication was set to 20 and the best particle position was migrated and mutated to substitute 25%, 50%, 75% and 100% of the poorer particles in the receiving group.

As shown in Table 8.2 and Table 8.3, the first communication strategy is effective for the parameters of solution that are independent or loosely correlated.

For the second experiment, the number of iterations for communication was set to 100 and the number of poorer particles substituted at each receiving

Table 8.2. Performance Comparison of *PSO* and *PPSO* with The First Communication Strategy for Rosenbrock Function

Percentage of	cost of $f_1(X)$			
Migration	PSO	PPSO(2,80)	PPSO(4,40)	PPSO(8,20)
None	108.739			
25%		65.384	98.548	75.948
50%		75.994	61.095	67.178
75%		61.190	64.507	59.955
100%		68.444	60.203	50.883

Table 8.3. Performance Comparison of *PSO* and *PPSO* with The First Communication Strategy for Rastrigin Function

Percentage of	cost of $f_2(X)$			
Migration	PSO	PPSO(2,80)	PPSO(4,40)	PPSO(8,20)
None	24.544			
25%		16.875	17.909	16.058
50%		15.123	12.875	12.835
75%		12.877	11.183	11.024
100%		11.243	10.507	10.030

Table 8.4. Performance Comparison of *PSO* and *PPSO* with The Second Communication Strategy for Griewank Function

Number of	cost of $f_3(X)$			
Migration	PSO	PPSO(2,80)	PPSO(4,40)	PPSO(8,20)
None	0.01191			
1		0.01137	0.00822	0.00404
2		0.01028	0.01004	0.00601

group was set to 1 and 2. Experimental results are shown in Table 8.4 and show that the second communication strategy for 8 groups may improve the performance by up to 66%.

For the third experiment, the parameters are the same as the first and second experiments. 50% of particles are substituted in the receiving group

Table 8.5. Performance Comparison of *PSO* and *PPSO* with The Third Communication Strategy

Function	cost		
	PSO	PPSO(4,40)	PPSO(8,20)
Rosenbrock	108.74	76.59	82.62
Rastrigin	24.54	18.63	19.14
Griewank	0.01191	0.01053	0.00989

for the first communication strategy and 2 particles are substituted in the receiving group for the second communication strategy. As shown in Table 8.5, the hybrid communication strategy can be effective for all three functions. The experimental results demonstrate the effectiveness and the computational efficiency of the proposed PPSO comparing with the traditional PSO.

8.4 Conclusions

In this work, a parallelised version of the particle swarm optimization scheme is presented. Three communication strategies for PPSO are discussed, which can be used according to the strength of the correlation of parameters. A third strategy is proposed in cases in which the characteristics of the parameters are unknown. Our experimental results demonstrate the usefulness of the parallel particle swarm optimization algorithm with these three communication strategies. Although the proposed PPSO may get the better performance, the optimal group size and the number of particles in each group should depend on the applications. It is the future work for designing adaptive group size and particle size in each group.

References

1. P. Angeline. Evolutionary optimization versus particle swarm optimization: philosophy and performance differences. In *Proc. Seventh Annual Conference on Evolutionary Programming*, pages 601–611, 1998.
2. J.-F. Chang, S. C. Chu, John F. Roddick, and J. S. Pan. A parallel particle swarm optimization algorithm with communication strategies. *Journal of Information Science and Engineering*, 21(4):809–818, 2005.
3. S. C. Chu, John F. Roddick, and J. S. Pan. Ant colony system with communication strategies. *Information Sciences*, 167:63–76, 2004.
4. M. Clerc and J. Kennedy. The particle swarm-explosion, stability, and convergence in a multidimensional complex space. *IEEE Transactions on Evolutionary Computation*, 6(1):58–73, 2002.

5. J. P. Cohoon, S. U. Hegde, W. N. Martine, and D. Richards. Punctuated equilibria: a parallel genetic algorithm. In *Second International Conference on Genetic Algorithms*, pages 148–154, 1987.
6. ed. Davis, L. *Handbook of genetic algorithms.* Van Nostrand Reinhold, New York, 1991.
7. R. Eberhart and X. Hu. Human tremor analysis using particle swarm optimization. In *Congress on Evolutionary Computation*, pages 1927–1930, 1999.
8. R. Eberhart and J. Kennedy. A new optimizer using particle swarm theory. In *Sixth International Symposium on Micro Machine and Human Science*, pages 39–43, 1995.
9. R. Eberhart and Y. Shi. Comparison between genetic algorithms and particle swarm optimization. In *Proc. Seventh Annual Conference on Evolutionary Programming*, pages 611–619, 1998.
10. Y. Fukuyama and H. Yoshida. A particle swarm optimization for reactive power and voltage control in electric power systems. In *Congress on Evolutionary Computation*, pages 87–93, 2001.
11. M. Gen and R. Cheng. *Genetic algorithm and engineering design.* John Wiley and Sons, New York, 1997.
12. D. E. Goldberg. *Genetic Algorithm in Search, Optimization and Machine Learning.* Addison-Wesley Publishing Company, 1989.
13. H. C. Huang, J. S. Pan, Z. M. Lu, S. H. Sun, and H. M. Hang. Vector quantization based on genetic simulated annealing. *Signal Processing*, 81(7):1513–1523, 2001.
14. J. Kennedy and R. Eberhart. Particle swarm optimization. In *IEEE International Conference on Neural Networks*, pages 1942–1948, 1995.
15. J. Kennedy and R. Eberhart. A discrete binary version of the particle swarm algorithm. In *IEEE International Conference on Computational Cybernetics and Simulation*, pages 4104–4108, 1997.
16. S. Kirkpatrick, Jr. C. D. Gelatt, and M. P. Vecchi. Optimization by simulated annealing. *Science*, 220(4598):671–680, 1983.
17. J. S. Pan, F. R. McInnes, and M. A. Jack. Application of parallel genetic algorithm and property of multiple global optima to vq codevector index assignment for noisy channels. *Electronics Letters*, 32(4):296–297, 1996.
18. Y. Shi and R. Eberhart. A modified particle swarm optimizer. In *IEEE World Congress on Computational Intelligence*, pages 69–73, 1998.
19. Y. Shi and R. Eberhart. Empirical study of particle swarm optimization. In *Congress on Evolutionary Computation*, pages 1945–1950, 1999.
20. Y. Shi and R. Eberhart. Fuzzy adaptive particle swarm optimization. In *Congress on Evolutionary Computation*, pages 101–106, 2001.
21. Peter Tarasewich and Patrick R. McMullen. Swarm intelligence. *Communications of the ACM*, 45(8):63–67, 2002.

9

Parallel Ant Colony Optimization for 3D Protein Structure Prediction using the HP Lattice Model

Daniel Chu and Albert Zomaya

School of IT, Faculty of Science,
University of Sydney, Australia
(dchu, zomaya)@it.usyd.edu.au, http://www.it.usyd.edu.au/ zomaya/

Protein structure prediction (also known as the protein folding problem) studies the way in which a protein will 'fold' into its natural state. Due to the enormous complexities involed in accuratly predicting protein structures, many simplifications have been proposed. The Hydrophobic-Hydrophilic (HP) method is one such method of simplifying the problem. In this chapter we introduce a novel method of solving the HP protein folding problem in both two and three dimensions using Ant Colony Optimizations and a distributed programming paradigm. Tests across a small number of processors indicate that the multiple colony distributed ACO (MACO) approach outperforms single colony implementations. Experimental results also demonstrate that the proposed algorithms perform well in terms of network scalability.

9.1 Introduction

Proteins [21] are complex molecules that consists of a chain of amino acids. They are essential to the structure and function of all living cells and perform functions such as acting as a catalyst for biomedical reactions, as structures of cells and receptors for hormones. By understanding how proteins achieve their native three dimensional structure, we can also help develop treatments for diseases such as Alziemers and Cystic Fibrosis.

While there have been intensive development in the field of protein structure prediction. It remains computationally infeasible to accurately predict structures of large proteins in any given condiction. The Hydrophobic-Hydrophilic (HP) lattice was developed to simplify the problem whilst maintaining behavioral relevance. Until recently the 3D HP model has been largely

D. Chu and A. Zomaya: *Parallel Ant Colony Optimization for 3D Protein Structure Prediction using the HP Lattice Model*, Studies in Computational Intelligence (SCI) **22**, 177–198 (2006)
www.springerlink.com

ignored in favour of the 2D HP model due to the simplicity of the search space and easy visualization.

The simplified HP models have been proven to be NP-Complete [4] thus providing a perfect platform for evaluating and improving heuristic based algorithms that will assist future development of expanded protein folding problems.

Much of the previous work in this field focused on developing and implementing solutions to 2d protein folding without considering 3D implementations. We propose a algorithm that can solve the 3D HP protein folding problem by extending a solution to the 2D HP problem.

9.2 Background

The Protein Folding Problem (or the Protein Structure Prediction Problem) is defined as the prediction of the 3D native structure of a given protein from its amino acid chain (primary structure). The native structure or state is the 3D structure that proteins vibrate around when equilibrium is formed almost immediately after protein creation. Experiments by Anfinsen et al in 1961 showed that a protein in it's natural environment will fold into its native state regardless of the starting confirmation.

The native structure of proteins is classically determined by techniques such as Nuclear-magnetic resonance imaging (NMRI) and xray/electron crystallography which are expensive in terms of computation, time and equipment.

9.2.1 Protein Structures

Proteins are polymer chains of amino acids. Each amino acid consists of a central carbon atom and four connecting bonds, namely a hydrogen atom, a carboxylic acid group, an amino group and a side chain. The carboxylic acid (COOH) and the amino group (NH_2) can dissociate or accept a hydrogen ion respectively, thus allowing the amino acid to act either as acid or base. The side chain also known as the R-Group acts as the distinguishing feature of the amino acid such that all twenty amino acids found in proteins have unique side chains.

During protein synthesis, amino acids are joined end-to-end by the formation of peptide bonds when the carboxyl group of one amino acid condenses with the amino group of the next to eliminate water. This process is repeated as the chain elongates. The amino group of the first amino acid and the carboxylic group of the last amino acid in the polypeptide chain remains unchanged and is named the amino terminus and carboxyl terminus respectively.

In biology, protein structures are usually described in four levels in order to reduce complexity. The amino acid sequence of a protein chain is called the primary structure. Different regions of the sequence form local secondary structures such as alpha helics and beta strands. The tertiary structure is

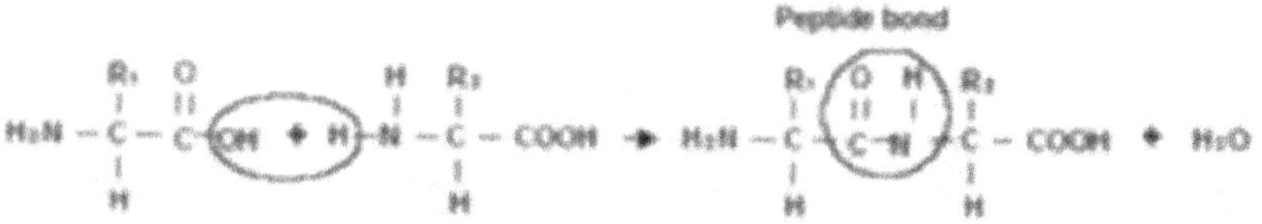

Fig. 9.1. Two amino acids forming a peptide chain

formed by packing such structure into one or several compact globular units called domains. Tertiary structure can also be regarded as the full 3D folded structure of the polypeptide chain. The final quaternary structure describes the interconnections and organization of multiple peptide chains and therefore is only used when there is more than one polypeptide chain. It is widely believed that secondary, tertiary and quaternary structures of the proteins native state can be determined by the primary structure.

9.2.2 De Novo Modelling

De novo or ab initio protein modelling methods seek to construct three-dimensional protein modells from the ground up. There are many possible procedures that either attempt to mimic protein folding (such as molecular dynamic simulations) or apply some stochastic method to search possible solutions. These procedures tend to require vast amounts of computational resources and are only used for smaller protein sequences.

9.2.3 Comparative Modelling

This technique utilises previously solved protein structures as a template. These methods generally fall into two classes

1. Homology modelling assumes two homologous proteins will share very similar structures. Given the amino acod sequence of an unknown structure and the solved structure of of a homologous protein, each amino acid in the solved structure is mutated computationally into the corresponding amino acid from the unkown structure.
2. Protein threading scans the amino acid sequence of an unknown structure against a database of solved structures. In each case, an evaluation function is used to assess the compatability of the sequence to the structure thus yeilding three dimensional models.

9.2.4 Molecular Dynamics Simulations

These simulations are a denovo modelling technique which attempts to mathematically solve Newton's equations of motion on an atomic scale. This method calculates the time dependant behaviour of a molecular system and is now routinely used to investigate the structure, dynamics and thermodynamics of molecular structures. Due to the compexities and high error rates involved in calculating inter-molecular and intra-molecular forces, a high level of tuning and adjustments are required to simulate a single protein. The computational costs involved also makes large-scale protein folding simulations unfeasible. Only large scale distributed systems such as the Folding@Home [22] experiment which uses thousands of distributed systems across a public network have broken the microsecond barrier using this technique.

9.2.5 Hydrophobic-Hydrophilic Lattice Model

The Hydrophobic-Hydrophilic (HP) Lattice Model [13,7] is a NP-Complete [4] free energy model that is motivate by a number of well known facts about the pivotal role of hydrophobic and polar amino-acids for protein structure -

1. Hydrophobic interaction is the driving force for protein folding and the hydrophobocity of amino acids is the main force for development of a native conformation of small globular proteins.
2. Native structures of many proteins are compact and have well-packed cores that are highly enriched in the hydrophobic residues as well as minimal solvent exposed non-polar surface areas [13,18].

In the HP model the amino acid sequence is abstracted to a sequence of hydrophobic (H) and hydrophilic (P) residues. The protein conformations of this sequence are then restricted to self-avoiding paths on a lattice; for our purpose the 2D HP model considered here consists of a 2D square lattice and the 3D HP model consists of a 3D cube lattice.

Based on the biological motivation given above , the energy of a conformation is defined as a number of topological contacts between hydrophobic amino-acids that are not neighbors in the given sequence. Specifically a conformation c with exactly n such contacts would have an energy value of $E(c) = n(-1)$.

The HP Protein Folding Problem can be formally defined as follows - Given an amino-acid sequence $s = s_1 s_2 s_3 .. s_n$, find an energy minimizing conformation of s, i.e., find $c^* \in C(s)$ such that $E^* = E(c^*) = min\{E(c) \mid c \in C\}$, where $C(s)$ is set of all valid conformations for s [12].

9.2.6 Existing Algorithms for the HP Lattice Model

Over the years several well known heuristic based optimization algorithms have been applied to solve protein folding in the HP lattice model. This includes Evolutionary algorithms (EA's) and Monte Carlo (MC) algorithms. Ant Colony optimizations have been used in 2D HP lattice models [19]. Tabu searching (Hill climbing optimizations) has been combined with GA's. Three dimensional lattice models have also been used with success. None of these

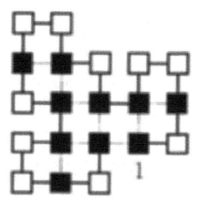

Fig. 9.2. A sample protein conformation in the 2D HP Model. The black squares represent Hydrophobic (H) residues. Dashed lines indicate the non-adjacent H residue contacts. The 1 indicates a terminating residue

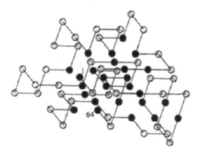

Fig. 9.3. A sample protein conformation in the 3D HP Model. The black squares represent Hydrophobic (H) residues. Dashed lines indicate the non-adjacent H residue contacts. The 1 indicates a terminating residue

solutions have have focused on distributed programming environments. With the assembly of computing systems with hundreds and even thousands of processors it is advantageous to harness this distributed processing power.

9.2.7 Evolutionary Algorithms

Evolutionary algorthms (EA's) cover a set of algorithms inspired by the theory of evolution. the most prominent of these algorithms are Genetic Algorithms (GA's). These techniques have been extensively used to solve the protein folding problem in the HP lattice model. Geneal priniciples for EA's folow the same principles - Solutions in the problem space are represented as a string. EA's construct an initail 'population' of solutions and in each iteration mutation and reproduction operators are probabalistically applied to 'evolve' a population over the lifetime of the experiment.

Unger and Moult [24] first presented an early application of EA's to the protein folding problem for both 2D and 3D HP models. Their method utilized an EA/ Monte carlo hybrid that initialized rge population with some valued solution. By using an absolute direction system (e.g. up is +1 along the z axis) they considered only valid conformations by repeatedly executing each operator until a valid conformation is achieved. The algorithm was able to generate high quality solutions for both 2D and 3D HP models. The test proteins of 64 amino acids in length used for their 3D implmentation is also used here.

Paton et al. [2] introduced a traditional genetic algorithm with a relative encoding scheme, invalid solutions are also accepted with a penalty value added. By using two point crossover and bit mutation better solutions were attained for all proteins with 64 amino acids other than Unger 64-1 and Unger 64-4

Khimasia and Coveney [12] focused on the design of a simple GA (SGA) for protein folding on the 3D lattice. Absoulte encoding, single point crossover and bit-wise mutation were used. Fitness proportional selection and elitism were also used. In these experiments invlaid conformations were also penalized. Results indicated that SGA underperformed in all of the Unger64 tests.

Aside from the work mentioned above, there are several other novel approached in using EA for the two dimensional model which may be of interest, in particular the combination of GA and tabu search proposed by T. Jiang et al. [11] and Evolutionary Monte Carlo (EMC) algorithm with secondary structures by Jiang et al. [14].

9.2.8 Hydrophobic Zipper (HZ)

The Hydrophobic Zipper (HZ) first proposed by Dill [1] is based on the observation that hetero-polymer collapse is driven by non-local interactions causing sheet and other irregular formations. Through small conformational searches the HZ algorithm attempts to move pairs of H monomers that are in relatively close proximity to each other and then use the newly formed contact as a search constraint for further pair formations. The movement in construction of the formations is similar to that of the action associated with the zipping of a zipper. In the 3D HP model the HZ algorithm found high quality solutions in excellent time. As only two test sequences (Dill-46 and Dill-58) were examined, algorithm performance for long protein chains remains inconclusive.

9.2.9 Contact Interaction

Contact interaction (CI) [23] is an extension of the standard Monte Carlo method which utilizes the concept of cooperation between non-local contact interactions. Rather than evaluating the molecule as a whole, H-H contact loops are examined. When combined with a cooling scheme the proposed algorithm discovered highly competitive solutions for all input test proteins. Due to the complexities involvced in handling loop detection and cooling, parallel implementations of such algorithm would be difficult at best and therefore not suitable for distributed implementation in a heterogenous computing environment.

9.3 Ant Colony Optimization

Ant Colony optimization (ACO) is a class of natural algorithms inspired by the foraging behavior of ant colonies. First proposed [15] as an approach to difficult optimization problems such as TSP [9], ACO has since been applied to many other problems [16][10][5][6]. The algorithm mimics the behavior of a colony of ants trying to find the shortest path between nest and food source by breaking down the problem into a shortest path problem and creating a colony of artificial ants to search for the shortest path (which is the optimal solution).

Algorithm 9.1 Single process ant colony algorithm

---
1.   initialize pheremone trails;
2.   while acceptable solution is not found do
3.   construct candidate solutions;
4.   perform local search;
5.   update pheremone trails;
6.   return best found solution;
end.

It is well known that biological ants form and maintain paths primarily through creation of a pheremone trail. Whilst traveling, ants deposit a certain number of pheremones and probabilistically chooses the direction richest with pheremone. Because shorter paths between the nest and food would be reached faster, the pheremones of the shorter path would also be stronger, resulting in other ants being more likely to choose the shorter path thus slowly eliminating the longer path.

The behavior of the artificial ants in ACO's are similar to that of biological ants. In each iteration of the algorithm each artificial ant acts as an agent traversing through the problem in order to construct a higher quality solution . At the end of each iteration the quality of all candidate solutions are evaluated based on a heuristic function and the top solutions are then employed to update the pheremone matrix.

In general ACO uses the following three ideas from natural ant behavior -

1. The preference for solutions or parts of solution with a higher pheremone level
2. The higher rate of growth of the amount of pheremone on better solutions
3. Trail mediated communication amongst ants

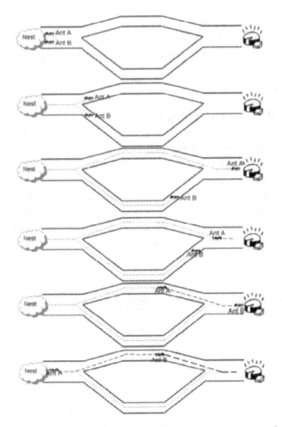

Fig. 9.4. Ants moving from nest to food todo – more explanation!

9.3.1 Pheremone Matrix

The pheremone matrix is how ACO keeps track of the pheremone values of given routes. All values within the matrix are initially set to zero, and the matrix is updated with pheremone values of the better paths with each iteration of the algorithm. The amount of pheremone added to the matrix is usually dependent on the quality of the ant, therefore proportional to the objective function value of a given solution. It is also quite common to multiply a constant p (where $0 < p < 1$) to the matrix every iteration to simulate evaporation of the pheremone trail when a path is not used.

9.3.2 Local Search

Many applications of ACO include a local search element as a means of by-passing local minima and preventing the algorithm converging too quickly. Usually algorithms utilized for local search are mutations of the acquired solutions or problem specific optimizations such as 2-opt for TSP.

9.3.3 Population Based ACO

Population ACO (PACO) [3] are designed to work in conjunction with other population algorithms such as GA and used when storing floating point values is either unfeasible or inefficient.

Rather than retaining a pheremone matrix at the end of the iteration, a population of solutions is kept instead. During the beginning of each iteration the population of solutions from previous iterations are used to construct the pheremone matrix which is then used to create the population at the next iteration. This allows the combination of ACO and other population based algorithms such as genetic and evolutionary algorithms.

9.3.4 Multi Colony ACO (MACO)

Multi colony algorithms (MACO) utilize multiple colonies of artificial ants. These solutions utilize separate pheremone matrices for each colony and allow limited cooperation between different colonies. MACOS are especially suited for distributed computing due to the limited amount of information exchange and synchronization required. Methods of information exchange include [17]:

1. Exchange of the global best solution every n iterations, where the best solution is broadcast to all colonies and becomes every colonies best local solution
2. Circular exchange of best solutions every n iterations. A virtual neighborhood is established such that colonies form a directed ring structure. Every n iterations every colony sends it's best local solution to the successor colony in the ring. The best local solution is updated accordingly.
3. Circular exchange of m best solutions every n iterations. The colonies form a virtual directed ring. Every n iterations every colony compares its m best ants with the m best ants of its successor in the ring. The best m ants are allowed to update the pheremone matrix.
4. Circular exchange of the best solution plus m best local solutions every n iterations.

9.4 ACO Implementation

We constructed four implementations of our 3D ACO – a reference implementation and three distributed implementations. The distributed models all use master/slave paradigms. The solution was assembled to report the number of cpu ticks that the program's master process took to find an improved solution as well as the score associated with the produced conformation.

9.4.1 Construction Phase

In the construction phase, each ant randomly selects a starting point within the given protein sequence. The sequence is then folded in both directions one amino acid at a time. The probability of extending the solution in each direction is equal to the number of unfolded amino acids in the respective direction divided by the total number of unfolded residues. This method encourages both directions of the chain completely folding within a few construction steps of one another.

In each construction step the relative direction is determined probabilistically using heuristic value $\eta_{i,d}$ and the pheremone values $\tau_{i,d}$ where d is the relative direction of folding at position i of the protein sequence.

Algorithm 9.2 ACO construction phase

---
1.   for each ant do;
2.   randomly select starting positions;
3.   while candidate solution not complete do
4.   select next amino acid;
5.   calculate possible directions and heuristics;
6.   if no possible direction found do backtrack
7.   else select next direction;
8.   return candidate solution;
end.

When the conformation is extended in the reverse direction (from i to $i-1$) the pheremone values $\tau'_{i,d}$ and heuristic $\eta'_{i,d}$ are used. In our implementation we define: $\tau'_{i,L} = \tau_{i,R}$, $\tau'_{i,R} = \tau_{i,L}$, $\tau'_{i,S} = \tau_{i,S}$, $\tau'_{i,U} = \tau_{i,U}$, $\tau'_{i,D} = \tau_{i,D}$ and analogous equalities for the respective η' values. This reflects the symmetry of traveling in opposite directions.

Once $\eta_{i,d}$ is calculated for the current amino acid the relative direction d of s_{i+1} with respect to s_i and s_{i-1} is determined according to

$$p_{i,d} = \frac{[\tau_{i,d}]^\alpha [\eta_{i,d}]^\beta}{\sum_{e \in \{L,R,S,U,D\}} [\tau_{i,e}]^\alpha [\eta_{i,e}]^\beta} \tag{9.1}$$

9.4.2 Heuristics

The heuristic value $\eta_{i,d}$ guides the construction process towards high quality solutions. In our implementation $\eta_{i,d}$ is defined as $\eta_{i,d} = h_{i+1,d} + 1$ where h is the number of new $H - H$ contacts achieved by placing s_{i+1} in the direction d relative to s_i and s_{i-1} when folding forward. Since only $H - H$ bonds contribute to the heuristics, $h_{i+1,d} = 0$ if the current amino acid is a P molecule and therefore $1 < i < n - 1$, $1 \leq \eta_{i,d} \leq 5$ and $1 \leq \eta_{n,d} \leq 6$.

9.4.3 Coding

As in other work [19] candidate conformations are represented through relative directions straight, left, right, up and down $(S, L, R, U and D)$ for the 3D lattice. Each direction in the candidate conformations indicates the position of the next amino acid relative to the direction projected from the previous to the current amino acid. The number of directions required for a conformation of size n is $n-2$. An orientation value is also required to determine the upward direction at a given amino acid. During the construction phase this indicator for the upward direction is stored.

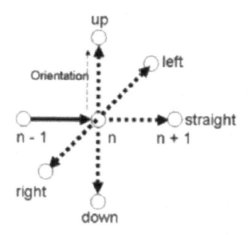

Fig. 9.5. Relative direction system for the candidate solution

9.4.4 Local Search

Local search is implemented similar to previous work [12] where we initially select a uniformly random position within a candidate solution and randomly change the direction of that particular amino acid.

9.4.5 Pheremone Updates

Selected ants update the pheremone values by the following -

$$\tau_{i,d} = (1 - \rho)\tau_{i,d} + E(c)/E^* .$$

ρ is the pheremone persistence that determines how much pheremone evaporates each iteration of the algorithm, $E(c)/E^*$ is the relative solution quality of the given candidate conformation where E^* is the known minimal energy for the given protein. If this value is unknown an approximation is calculated by counting the number of H residues in the sequence. This technique prevents stagnation of solutions and ensures that lesser quality candidate solutions contribute proportionally lower amounts of pheremone.

Algorithm 9.3 ACO local search

1. copy c to c';
2. choose sequence index i randomly from 1 .. n;
3. randomly change d at index i of conformation c dash;
4. while c' is invalid do
5.   find invalid direction;
6.   change invalid direction based on heuristic from c;
7. return minimun(E(c dash), E(c));
end.

9.4.6 Single Process Single Colony

The reference implementation utilises a single processor, single colony and single pheromone matrix. Every distributed implementation would function in this fashion should there be only one processor avaliable.

9.4.7 Distributed Single Colony

In this implementation, all slave systems share the same master pheromone matrix. Each slave initally behaves as a independent colonly until the end of construction and local search phases. During the pheromone update phase, rather than updating the local pheromone matrix, selected ants are transmitted to the master to update the global matrix. Subsequently a updated version of the global matrix which during this time may have been updated by other slaves is transmitted to the slave to be used in the next iteration of the algorithm

---

Algorithm 9.4 DSC server algorithm

1. initialize n number of pheremone matrices;
2. while termination-condition not met do
3.   while candidate solution not complete do
4.   wait until a client solution is recieved;
5.   update pheremone matrix;
6.   transmit to client updated pheremone matrix;
end.

9.4.8 Distributed Multi Colony with Circular Exchange of Migrants

This multi-colony implemenation assumes each slave is an independent colony. Based on the theory that some pheromone left behind by different colonies may have a limited effect on each other. The pheremone matrices of each colony remain on the master for crash recover. in every iteration, client transmits selected conformations at the end of local search phase for pheremone updates and receives a updated pheremone matrix. In every n iterations for each colony, the pheremone matrix of the given colony is updated by a neighbour such that they form a ring shape.

Algorithm 9.5 MCOCE server algorithm

---
1.   initialize n number of pheremone matrices;
2.   initialize n number of integer counters;
3.   initialize pheremone matrix ring;
4.   while (termination-condition not met) do
5.   wait until a client sends their solution;
6.   update client's respective pheremone matrix;
7.   increment this client's counter;
8.   if (counter gt iteration limit) do
9.   reset counter;
10.   update clients neighbour's pheremone matrix;
11.   set marker for client neighbour to receive updated matrix;
end.

9.4.9 Distributed Mulit Colony with Pheremone Matrix Sharing

Each slave process in this implementation is an independent colony, at every n total iterations counted on the master, every pheremones matrix is updated by the following formula:

$$tau_{C,i,d} = \tau_{C,i,d} + \frac{(\sum_{n=1}^{Ctotal} \tau_{n,i,d} - \tau_{C,i,d})\gamma}{Ctotal - 1} \tag{9.2}$$

9.5 Results

9.5.1 Testing Methodology

For comprehensive evaluation of results, inputs from three well known sources of examples for the 3D HP problem [1, 24, 23, 20, 8] was drawn upon. For comparison between various distributed implementations, the following test set up was implemented

---

Algorithm 9.6 MCOPE server algorithm

---

1. initialize n number of pheremone matrices;
2. initialize counter;
3. while (termination-condition not met) do
4.   wait until a client sends their solution;
5.   update client's respective pheremone matrix;
6.   increment this client's counter;
7.   if(counter gt iteration limit * number of colonies) do
8.   reset counter;
9.   update all pheremone matrices;
10.   update all clients with updated pheremone matrices;
end.

---

1. All ant colonies are set to a size of 1500
2. From testing of the single colony ACO algorithm (something) values are all set to 1, 2 and 0.2 respectively
3. Pheromone update frequency for distributed algorithms set to 15
4. value is set to 0.1 for MCOPE (experiments during parameter tuning indicate lower (something) values of increases convergence speed with no noticable loss in solution quality)
5. All tests consists of 10 consecutive runs and limited to 5 minutes each

Tests were also run on smaller protein chains taken from Unger and Moult [24]. The optimal solutions to these sequences was found in less than a minute. Most existing work can find these optimal soulutions and the data relating to these tests is not included.

9.5.2 Emperical Results

Due to the small convergence time a sample input protein of length 136 was used for the follwing graphs. The median values at given time intervals were calculated from 10 seperate runs and limited to 10 minutes each.
    sectionAnalysis

9.5.3 Convergence Speed Differences

Additional CPUs contribute to a faster convergence speed and in most cases lead to improved solutions being found. As expected the DSC exhibited linear convergance as the number of CPUs increased without improving on the best solution found. Due to the small percentage increase realized by using the other distributed techniques DSC is recommended when reasonable quality solutions are required in a short period of time.

    The MCOPE Algorithm provides the fastest initial improvemnet in solution quality of all the implementations followed by a slow down. This behaviour can be attributed to equalization methods that increase the size of

Table 9.1. Test sequences

Sequence name	amino acid string
Unger64-1	pphhhhhppphhppppphhpppphpppppphphppphpphphppppph ppppphhphhpphpphp
Unger64-2	pphphpphpphhhhphhhhpphhhpppphphppphphppphphppppph phpphphppphpphpp
Unger64-3	hphhpphhphpppppphhhphhhhpphphphphhpppphphpphhhphhph ppppphhhhhhhhppp
Unger64-4	hpphhpphpphphphpphppppphpppppphphphhhpphphpppphphpph hpphpphphphphhph
Unger64-5	hppphhpphphpppphppphphhpppphhphphhphpppphpppphpphphhh pphpphpphhhphhhh
Unger64-6	hpphhphhhhpppppphhpphpppphhpppphpphphhphpppphhppp phpppppphpppphphh
unger64-7	ppppphpppphpppphhhhphhpppppphpphphhphphphpppppphpppppppp ppphhhhppppphhpph
Unger64-8	ppphhhpphphphpphpphhpppphpphphphhphpppphpppppppphphhhp hhhhhpphhpphphpph
Unger64-9	hpphpphhhpppphphpppphphhphhhhhpppphphphpppphphpppp hhphpppphpphhphp
Unger64-10	pphpphpphhhpppphphpphpphpppppphpphhhpphpphpphphpp ppppphhhpppppphphp
Dill-46	phhhphhhppphphphphhpphphhhhphphhhhhhphphhpphhp
Dill-58	hphhhphhhpphphphphhphhhphphphhpphhhpphphppppphpphp phhpphpp
TOMA-136	hppppphpppphphhphhppppphphhhppppphphphhhhpppppppppp pphpphpppphphhpppphhpphpphphphppppppppppphppphhhhhhpp phhpphpphhhpphphpppppphhhhhhhhhhphhhhhphhhhh
TOMA-124	pphhhphpppphpppppphhpppphhpphhpppphppppphpphpphhpp phhphphhhppppphhhppppppphhpphpphphpphppppppphpphhh ppppphpppphhhhhppppphhphphphph

the search space in an effort to slow down convergance. The first instances of equlaization would most likely be detremental as the low values stored in the pheremone matrix at this time imply a large number of poor performing solutions being shared. The quality of solutions improves later in the experiment as a result of these solutions being removed from the matrix as the corresponding pheremone matrix evaporates.

9.5.4 Scalability

Two seperate tests were devised to evaluate the performace of the algorithms relative to the protein length. The first was the number of seconds to execute an iteration of the ACO algorithm for proteins of length 32-3000. The second test measures the number of iterations of the ACO algorithm required for

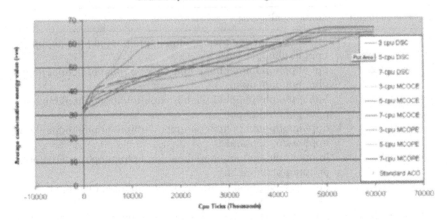

Fig. 9.6. Speed differences between applications

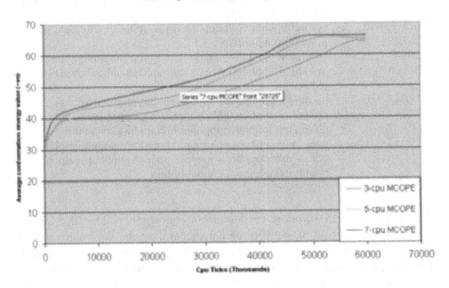

Fig. 9.7. Trends in the MCOPE algorithm

Table 9.2. Emperical results

sequence	old E*	aco	3-sc	5-sc	7-sc	3-mc	5-mc	7-mc	3-pe	5-pe	7-pe
Unger64-1	-27	-27	-27	-28	-30	-28	-29	-30	-29	-29	-30
Unger64-2	-30	-31	-34	-34	-34	-35	-33	-34	-34	-35	-35
Unger64-3	-38	-38	-41	-42	-42	-43	-43	-43	-44	-44	-44
Unger64-4	-34	-35	-36	-37	-37	-36	-38	-38	-37	-37	-37
Unger64-5	-36	-35	-37	-39	-39	-38	-40	-40	-40	-40	-40
Unger64-6	-31	-31	-31	-32	-32	-32	-31	-32	-31	-31	-32
Unger64-7	-25	-25	-26	-26	-26	-26	-27	-27	-26	-26	-26
Unger64-8	-34	-34	-34	-34	-35	-34	-35	-35	-34	-35	-35
Unger64-9	-33	-33	-35	-35	-36	-36	-38	-38	-36	-38	-38
Unger64-10	-26	-28	-29	-30	-30	-29	-30	-30	-29	-30	-31
Dill-46	-34	-34	-34	-34	-34	-34	-34	-34	-34	-34	-34
Dill-58	-42	-42	-42	-42	-42	-42	-43	-43	-42	-43	-43
Toma-136*	-65	-65	-65	-66	-66	-65	-67	-67	-66	-67	-68
Toma-124*	-58	-58	-59	-59	-59	-59	-59	-60	-59	-59	-60

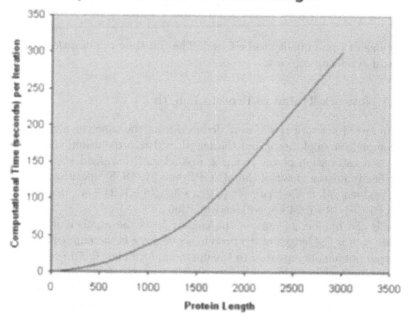

Fig. 9.8. Computation time per iteration versus protein length

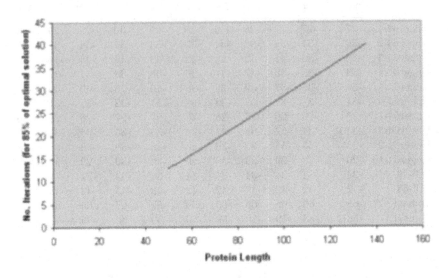

Fig. 9.9. Number of iterations versus protein length

solutions of equal quality to be found. The run time per iteration test reveal a nlogn sytle runtime cost.

9.5.5 Network Traffic vs Protein Length

Estimates of network traffic were deduced from the time the algorithm takes to complete a single iteration, the length of the protein and the number of ants. An estimation of the required upload and download speed was made specifically for the client of the MCOPE and MCOCE algorithms:

/beginequation $upstream = (0.01a * 3n)/(8 * 1024 * b_n)$ $downstream = (128 * 5 * n)/(8 * 1024 * b_n * c)$/endequation

a is the number of ants in the colony, b_n is the number of seconds per iteration, n is the length of the protein sequence, c is the number of iterations between pheremone updates. In the upstream formulae, $0.001a$ represents the top one percent of candidate solutions that are transfered; $3n$ is the amount of bits required for each candidate solution. In the downstream case, $128*5*n$ is the required number of bits requiried to transmit a pheremone matrix $5*n$ where n is the protein length.

The above formulae assume a sanitized heterogenous environment where all processors execute at approximately the same speed. In a heterogenous environment with resource contention, required upstream/downstream speed may differ.

In the case of the DSC algorithm, the upstream requirement would increase byt a factor of c due to the necessity to update the client's pheremone matrix every iteration. As expected an increase in protein length decreases the required rate of transmission given the low bandwith requirements both network and algorithm performance degredation would be negligible – even systems equiped with a 56kbps modem would be able to function as a computational client working on larger protein chain.

Network Traffic vs Protein Length For Clients

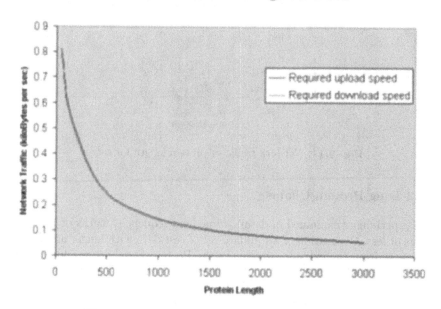

Fig. 9.10. Network traffic vs protein length for client

9.6 Conclusion and Future Work

9.6.1 Solution Quality

The single processor solution would not find an optimal solution in all cases. This was to be expected considering the small search space that the algorithm covers. In these cases we stopped the solution once the most optimal solution was found. Both multiple colony implementations out performed the single colony implementation across 5 processors by a large margin. The tests run across five processors for multiple colonies resulted in execution times of less than a second – we did not test on more processors in these cases.

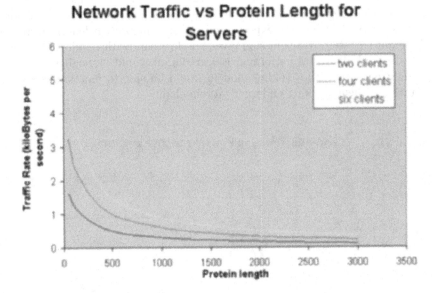

Fig. 9.11. Network traffic vs protein length for server

9.6.2 Long Protein Chains

The experiments examined in detail amino acid sequence lengths of 136 characters or less. Generation of a standard set of test data with longer amino acid sequences from the existing proteins in the Protein Database would be benificial as this will provide a universal benchmark for which new and existing alorithms may be tested against.

9.6.3 Variable Colony Sizes

ne possible improvent to the ACO is the introduction of variable colony sizes. Currently all proposed implementations of ACO utilise a static numbers of ants. It is possible that by adjusting the number of ants according to the rate of convergance, computational speed or network requirements an improvemnt in the quality of solutions found may yet be seen.

9.6.4 Investigation of Fitness Landscape

It has been noted that proteins may have several native conformations depending on environmental variables such as temperature. This suggests mutiple global optimal may exist in the fitness landscape and maybe of interest to further investigate the fitness landscape of the 3D HO model and how the landscape may correlate to real proteins. A more advanced model may then be introduced from which more knowledge regarding protein folding may be deduced.

We have developed a framework for testing a variable number of colonies for ACO's for the HP protein folding problem. We have shown that good 2D solutions for this problem can be extended to the 3D case. Our solution works well across a number of processors. Multiple colony solutions offer a definite improvement over the single colony implementation. We hope to harness other properties of ACO's by extending our solution to work across loosely coupled distributed systems such as grids.

References

1. Klaus M. Fiebig A. Dill and Hue Sun Chan. Cooperativity in protein folding kinetics. Proceedings of the National Academy of Sciences, 90:1942-1946, 1993.
2. W. Punch A. Patton and E. Goodman. A standard GA approach to native protein conformation prediction. Proc Incl Conf on Genetic Algorithms, pages 574-581, 1995.
3. M. Guntsch M. Middendorf O. Diessel H. ElGindy H. Schmeck B. Scheuermann, K.So. Applied Soft Computing 4, pages 303-322, 2004.
4. Bonnie Berger and Tom Leighton. Protein folding in the hydrophobic – hydrophilic (HP) model is np-complete. Journal of Computational Biology, 5(1):27-40, 1998.
5. R.F. Hartl Bullnheimer D and C Strauss. Ant Colony Optimization in Vehicle Routing. PhD thesis, University of Vienna, 1999.
6. Costa D. and A. Hertz. Ants can Colour Graphs. Journal of Operational Research Society, 48:295-305, 1997.
7. K.A. Dill. Biochemistry 1985.
8. J. Flores, S.D. Smith. The 2003 Congress on Evoluntionary Computation, 4:2338-2345, 2003.
9. Dorigo M.L.M Gambardella. Ant Colonies for the Travelling Salesman Problem. BioSystems, 43:73-81, 1997.
10. E. Taillard Gambardella L.M and M. Dorigo. Ant colonies for the quadratic assignment problem. Journal of the Operational Research Society, 50:167-176, 1999.
11. T.Z. Jiang. Protein folding simulations of the hydrophobic-hydrophilic model by combining tabu search with genetic algorithms. The Journal of Chemical Physics 119(8):4592, Aug 2003.
12. M. Khimasia and P. Coveney. Protein structure prediction as a hard optimization problem: the genetic algorithm approach. Molecular Simulation, 19:205-226, 1997.
13. K.F. Lau and K.A. Dill. A lattice statistical mechanics model of the conformation and sequence space for proteins. Macromolecules, 22, 1989.
14. Faming Liang and Wing Hung Wong. Evolutionary mote carlo for protein folding simulations. The Journal of Chemical Physics, 115(7):3374-3380, August 2001.
15. Dorigo M. Optimization, Learning and Natural Algorithms. PhD thesis, Politicnico di Milano, 1992.

198   Daniel Chu and Albert Zomaya

16. Gambardella L.M and M. Dorigo. Has-sop: An hybrid ant system for the sequential ordering problem. Technical Report IDSIA 97-11, Lugano, Switzerland, 1997

17. H. Schmek M.Middendorf, F. Reischle. Information exchange in multi colony ant algorithms. Parallel and Distributed Computing: Proceedings in the Fifteenth IPDPS Workshops 2000, 2000

18. F.M Richards. Areas, volumes, packing and protein structures. Annu. Rev. Biophys. Bioeng., 6:151-176, 1977.

19. A Shmygelska and H.H. Hoos. An improved ant colony optimization algorithm for the 2d HP protein folding problem. Canadian Conference on AI 2003, pages 400-417, 2003.

20. W. Hart, S. Istrail. HP Benchmarks http://www.cs.sandia.gov/tech_reports/compbio/tortilla-hp-benchmarks.html last accessed October 2004

21. Branden. C, Tooze. J, Introduction to Protein Structure. Garland Publ, New York, London.

22. Michael Shirts Stefan M. Larson, Christopher D.Snow and Vijay S.Pande. Using distributed computing to tackle previously intractable problems in computational biology. 2002.

23. L.TOMO and A.TOMA. Contact interactions method: A new algorithm for protein folding simulations. Protein Science, 5(1):147-153, 1996.

24. Ron Unger and John Moult. A genetic algorithm for 3d protein folding simulations. Proceedings of the Fifth Annual International Conference on Genetic Algorithms, pages 581-588, 1993.

Subject Index

Author Index